Diddly Squat

Pigs Might Fly

Diddly Squat

Pigs Might Fly

JEREMY CLARKSON

MICHAEL JOSEPH

PENGUIN MICHAEL JOSEPH

UK | USA | Canada | Ireland | Australia
India | New Zealand | South Africa

Penguin Michael Joseph is part of the Penguin Random House group of companies
whose addresses can be found at global.penguinrandomhouse.com

First published 2023

004

Copyright © Jeremy Clarkson, 2022, 2023

The moral right of the author has been asserted

Set in 14.71/19.62pt Garamond MT Std
Typeset by Jouve (UK), Milton Keynes
Printed and bound by CPI Group (UK) Ltd, Croydon CR0 4YY

The authorized representative in the EEA is Penguin Random House Ireland,
Morrison Chambers, 32 Nassau Street, Dublin D02 YH68

A CIP catalogue record for this book is available from the British Library

HARDBACK ISBN: 978–0–241–67489–5
TRADE PAPERBACK ISBN: 978–0–241–67490–1

www.greenpenguin.co.uk

Penguin Random House is committed to a
sustainable future for our business, our readers
and our planet. This book is made from Forest
Stewardship Council® certified paper.

Contents

SUMMER

Green-winged testicles   3

Break-Heart Maestro   11

Give my 'little man' a call   19

AUTUMN

Three from the sun   29

True heaven is a place on earth   37

A twitcher's lament   45

David Cameron murdered my tractor   53

I've become an eco-farmer – and Kaleb
      hates it   61

WINTER

(Lisa is) Armed and dangerous   73

CONTENTS

Bleak House   81

Another fine mess   89

Now I am become the Destroyer of Worlds   97

Walking the dog   105

SPRING

Environmental doughnuts   117

A thing that seems very thingish   125

A total pig's ear of it   133

Lady garden   141

I can't explain   149

SUMMER

Let them eat soup   159

In praise of goats   167

The contents of this book first appeared in Jeremy Clarkson's *Sunday Times* column. Read more about the world according to Clarkson every week in the *Sunday Times*.

SUMMER

Green-winged testicles

I have always been very suspicious of people who have hobbies. Because a hobby carves a hole in your soul and diminishes your humanity. It turns you into a monoculture, capable of speaking about and doing only one thing. Some people live to work and some work to live. A hobbyist works and lives to collect stamps. It's a dangerous thing.

I know this because, back in the early Seventies, Esso launched a marketing scheme where customers could collect a set of footballing coins. You got a coin with every four gallons you bought and I became so consumed with this, I used to make my dad drive everywhere in third so he'd use more petrol.

He was a travelling salesman at the time and had an account with Blue Star garages, but the only Blue Star garage that sold Esso petrol was on the Finchley Road in London. 'Well, can you fill up there?' I'd say tearfully. 'Not really, because we live in Doncaster.'

My sister had the notoriously difficult to find Kilmarnock coin and I seriously considered murdering her for it. I also considered murdering schoolfriends who wouldn't do swapsies. This is why I know that hobbies are dangerous.

Look at golf. One day you slip into a pair of Rupert Bear trousers and head off for a game with some friends. It'll seem pretty harmless, but sooner or later you're going to hit the ball in such a way that it goes in vaguely the right direction. And that'll be that. Next thing you know you'll be so consumed with the need to improve, you'll be spending all your money on better bats and you'll practise constantly, in the rain and on your own, until eventually your wife will leave you. And you won't notice for three months.

It's the same story with light aircraft enthusiasts and internet onanists. Hobbies become all-consuming. One minute you're catching sticklebacks on a lovely summer's day in the Test and the next you're friendless and damp on the banks of a terrible canal in Birmingham.

We are now starting to see the same problem with people who collect orchids. It starts harmlessly, when someone buys you a pretty example as a house-warming present and it ends with you on the dark web, at three in the morning, trying to find a man who'll go into the fields at night and steal you a variety that's on the edge of extinction.

I wish I was joking, but in recent months gardeners

and landowners in Kent and Sussex have been reporting a spate of thefts from their wildflower meadows. They go to bed at night, happy that a rare lady's slipper orchid, worth about £2,000 to the hobbyists, is growing on their land, and they wake in the morning to find nothing but a hole in the ground.

It's not only a profitable crime but easy too. Unlike Range Rovers, orchids don't have steering locks and alarms. Furthermore, if you're caught, and that is highly unlikely, the penalties are likely to be less severe. Yes, section 13 of the Wildlife and Countryside Act 1981 provides for a fine of £5,000 and six months in jail, but that's never going to happen because, at the end of the day, all you were actually doing, m'lud, is picking wild-flowers. And people have been doing that for years.

Back in 1956 someone dug up an orchid called summer lady's tresses from a site in the New Forest, and as a result it's now extinct. Thieves stole what was thought at the time to be the last lady's slipper orchid in 1917. And collectors damn nearly wiped out the lizard orchid. They must have known, because even I know it, that you can't dig up an orchid and expect it to thrive in your garden or on your hall table. Because it won't. It'll die.

I suppose it's the hobbyist egg collector mentality that was all the rage in Victorian times. Boffins would announce that there was only one great crusted lesser-spotted dodo left in the world and there'd be a stampede of hunting fanatics, each one desperate to kill it. 'Yes! I single-handedly wiped out an entire species. I ended it. And then I cooked its egg and ate that.'

And it's still going on today. There are orchid collectors who are so determined to have the equivalent of that Kilmarnock football coin that they'll stop at nothing to get it. That's why, in Yorkshire, there's one very rare orchid that is housed in a metal cage and monitored round the clock by CCTV. If someone could nick that, it'd probably make a Hollywood movie. Except *Adaptation*, a critically acclaimed film starring Nicolas Cage and Meryl Streep, of course got there first.

There are orchids on my farm. And some are very rare. I know this because when they were found by members of a local horticulturalist society they all ran around clutching their tinkles. It worries me that someone will steal them.

And it's not just orchids that are being pinched. Thieves are also targeting bluebells and, back in 2019, 13,000 snowdrops worth £1,500 were nicked from woodland in Norfolk. They're even going after ferns for God's sake. And now I'm wondering, how long will it be before thieves start to come after my wheat? Because thanks to the cost of fertiliser, one ear alone is now more valuable than a Rwandan water lily.

My big worry, though, is that topsoil could well become the next big thing for Johnny Burglar. Because it was recently described by the author Claire Ratinon as a 'divine entity'.

I was in Hampshire recently and as I drove along I became consumed with envy at the lushness of everything. Plainly the Meon Valley has much better soil than the stony brash I have up here in the Cotswold Hills. So why don't I just nick some? Is it even illegal, I wonder.

Certainly the money's good. It would take a skilled operator no more than 15 minutes to fill a truck, and that load would be worth maybe £2,000. So in a single night I could earn 50 grand.

There's only one thing stopping me really. My JCB telehandler. The first time I used it three years ago, I thought, 'Hang on. This is fun.' And the next step on that road is that it becomes a hobby. Already my relationship with it is worrying because I'm forever thinking of jobs that don't really need doing, just so I can play with it. If I thought it could earn me £50,000 a night, you'd never get me out of it.

Only the other day I was sating the needs of my inner nine-year-old by making an unnecessary bund out of some subsoil when a neighbour walked by and said I was operating perilously close to a clump of green-winged orchids.

I got out for a closer look and, I must say, they were very unimpressive. But then it was pointed out to me that orchid is derived from the Greek word *orchis*. Which means testicle. Yes. I have green-winged testicles on my farm. And I'm going to make it my life's work to look after them.

Break-Heart Maestro

I've rented a bull. He's the same shape as a mansion house. He has testicles the size of space hoppers. And his name, magnificently, is Break-Heart Maestro.

I wasn't going to use a bull. I've been trying to make my cows pregnant by putting them in a cramped space and deploying a turkey baster. Well, Boris Becker's short-lived girlfriend supposedly got pregnant by him in a broom cupboard at Nobu. But sadly, it wasn't working for Pepper, my all-white heifer.

The vet has been shoulder deep in her back bottom on a number of occasions and he assures me that if you have a good feel around beyond her cervix, she is sporting all the flobbery bits necessary to make baby cowlets. But even though Tim the cow man had injected her three times with a straw full of semen, she remains studiously unpregnant.

Hence the bull. And since he was coming, I figured he may as well use those hairy space hoppers to fill up all the other lady cows as well. With a bit of luck, he'd get them all up the duff at roughly the same time, which would make calving easier, and shorter, in nine months' time. Yup, as *Countryfile*'s Adam Henson said recently, cows have the same gestation period as 'people'. I know he meant 'women' but he's on the BBC, so he couldn't say that.

I woke early on the day the bull was due to arrive and I'll be honest, I was a bit worried. I'm not frightened of much. I'll fish a spider out of the bath and on a recent trip to the Seychelles, I swam after a five-foot blacktip shark to get a better look at it. But bulls worry me.

As they are basically life-support systems for their own testes, they are a tonne and a half of pure, undiluted testosterone. The bull I'd rented was four years old, which makes him, in human terms, like a teenage boy. If he could drive, he'd do so at breakneck speed. And if someone looked at him in a pub, he'd go at them fists flailing.

Regularly, in the farming magazines, there are tales of agricultural workers who've been mauled or trampled by bulls. And these are people who know what they're doing. Me? I can deliver a piece to camera while driving round a corner at 150mph but when it comes to man cows, I'm absolutely at a loss.

I did know, however, that before he arrived, I'd have to put up signs warning people on the public footpath that there was a meat machine in the adjoining field. So

I had some made saying 'Bull. Best not wear red trou-sers'. I thought this veiled dig at the awful people in the village who continue to make my life difficult was funny, but the experts disagreed. They said that by suggesting red trousers are inappropriate, someone in blue trousers may think he's free to leap over the fence.

I therefore had some other, straighter signs made saying 'Danger. Bull in field'. But this is also a no-no, as I'd be admitting that I know the bull is dangerous. Which would cause a QC to stand up in court and say, with all the mock incredulity he could muster, 'So you're saying, Mr Clarkson, that you put an animal that you knew to be dangerous, in a field near a footpath.' And then I'd have to go to prison for a thousand years.

In the end I settled on a simple sign that says 'Bull in field', and as I attached it to the gate, Break-Heart Maestro arrived.

He wasn't particularly enormous and, as a shorthorn, he didn't come with Mad Max antlers on his head. But you could just sense that he was extremely strong. I looked at the hedge cutter I'd used to block one possible means of escape and thought, 'He could flick that out of the way without even blinking.'

I don't hold with bullfighting. I think it's unnecessary and cruel. But as I stared into Break-Heart's dead eyes, I did feel a surge of respect for those Spaniards who stand in front of an enraged bull with nothing for protection apart from a red bedsheet. I wouldn't.

As I pondered this, my cows came round the corner. They gave birth only a few weeks ago and some were still breastfeeding. (It's not called that on a farm, by the way.) But immediately, their walk turned into a sashay. And I swear they were fluttering their eyelashes. Then the competition began.

Genghis the attack cow bolted first. Realising that she's far from the prettiest in the flock, she thought she'd better get in early, and broke into an undignified run. This caused the others to get a move on too, and pretty soon, we had a scene like something from an episode of *Bonanza*.

I was immediately worried, because what if Break-Heart decided to go for one of the calves? He'd snap it. But I was forgetting. Unlike men (or people in BBC-speak), bulls are only up for sex when the lady cows are in season. And the calves weren't old enough yet.

Besides, Break-Heart didn't appear to be up for sex with anything. Unlike Wayne Rooney, Leonardo DiCaprio and my rams, he didn't get on with the job straight away. He was playing it cool. Pretending not to be interested. He chewed the grass in his temporary new home and took in the views. Until, after a minute or so, one of the cows could wait no more and mounted him. She was thrusting away like a good 'un and what made this peg-ging scene even more surreal is that her two-month-old calf was looking on.

That evening, things got even more pornographical. Pepper, the white and unpregnant heifer, was standing on her own, when Break-Heart strolled over. 'Yes!!!' I thought. This is your moment, girl. But just as the mighty bull was positioning himself, another cow reversed over to him and issued a mighty stream of urine. He liked that so much that he spent the next ten minutes gently licking her back end. Pepper walked off, a sad look on her face.

He has been on the farm for four days now and, so far, I've seen a lot of foreplay but no actual mating. Appar-ently, it could take a couple of months for him to do the lot. And then he'll have to be tested for any diseases he's picked up while here before moving on to the next flock. Not a bad job really.

Of course, before that happens, I've got to make sure he doesn't kill me, and I've worked out a way this can be achieved. I'm going to film a new *Grand Tour* special and ask Kaleb to do all the cowboy stuff.

I realise there's still the problem of ramblers getting mauled. But hey-ho. Worse things happen.

Give my 'little man' a call

Now that we know it'll cost eleventy million pounds to heat our houses this winter, and that we won't be able to afford to pay because our monthly mortgage repayments will be bigger than the gross domestic product of Switzerland, many commentators are offering handy tips on how to save money.

You should put on an extra jumper, they say, and burn litter if you want to stay warm. You should glue your teeth together so you don't waste money on food and if you do, by some miracle, need the lavatory, you should wipe your bottom with a smooth stone to save on bog roll. It all sounds very uncomfortable and horrid. But happily, I have an easy and actually quite pleasant way of saving yourself an absolute fortune. Simply remove from your life every single luxury brand name.

Three or four years ago, when the world was joined up by ships and planes and everything worked, I had an hour to kill in London before lunch, so Lisa suggested we go into a shop called Chanel to buy a frock. Like all frocks, it was made from a bit of cotton thread and the shavings from a sheep, which will have cost the manu-facturer, at most, 15p. But because the label said 'Chanel' it cost 2,000 times more than a frock, made from exactly the same materials, that you could buy on a market stall in Bradford.

And what about pens? You could have something made by Montblanc which will cost you £2,400 but will it be any better at jotting stuff down than a Bic biro, which will cost you 10p? It's the same story with watches, suitcases, shoes, cars, jewellery or indeed anything you can buy in those awful 'brand' shops that litter airport departure lounges and high-end holiday resorts.

And then there's furniture. I recently built a house and needed a sort of sideboard thing for the hall. So I went to all the trendy places in Chelsea and Notting Hill and everything cost more than most tropical islands. And nothing could be delivered until the third quarter of 2028. I therefore decided an antique alternative would be better and started trawling various shops in the Cotswolds.

God, there are some bargains to be had – and antiques, because they were made in the 18th century, are available now. However, it's tricky to find exactly what you want. We spent weeks wandering around Stow-on-the-Wold and Chipping Norton and the giant bric-a-brac hangars outside Malmesbury and everything was either the right size and a bit ugly, or beautiful but the wrong shape.

Then we heard about a 'little man' in a nearby village. An old-school craftsman. A chap who wears dungarees and gets away with it. He operates out of what's nothing more than a hut, which is rammed full of tables he's restoring and chairs he's reupholstering and yes, he could make us a sideboard that would fill, exactly, the 12ft gap between the loo door and the door to the kitchen.

And now he has, and it's one of the most exquisite and beautifully crafted things I've ever seen. It's made from Brazilian rosewood, has a splash of marquetry, legs more captivating than Elle Macpherson's and lovely lion head handles. And the cost? It would be rude to say here but it's way shy of anything modern that you could find in London.

Buoyed by this, we found another 'little man' in Cirencester who, for a fraction of what we'd pay if we went to a designer brand, could make us a sofa to fit, precisely, in our new TV room. And he'd upholster it in any material that took our fancy and stuff it with filo-plume feathers from a red-tailed hawk if that's what we wanted. I was so pleased with the result, I've now asked him to make another sofa for the sitting room. He's making it out of myrrh and stuffing it with a bag of downy angel pubes, and I shall pay for it using the small change in the door pocket of my car.

This brings me on to wallpaper. The choice you have

from existing brands is bewildering. So bewildering that it would be impossible for the average couple to agree on anything. Amazingly, Lisa and I agreed that the downstairs loo should be lined with black flock wallpaper but, having decided this, we were presented with several million alternatives, which caused massive rows, many slammed doors, separate beds for about a month and much blocking on Instagram.

So we decided to abandon the established brands and chose instead to have our wallpaper handpainted by a 'little man' from the Middle Ages. I can't remember where he lives. Some place where they believe in ley lines. Somerset? Wiltshire?

Whatever, he's not being particularly speedy and nor will he show us what he's actually doing. This is the way with 'little men' in the village. They're craftsmen, not businessmen, and in their mind, the customer isn't just wrong; he's a damn nuisance.

But, after a six-month wait, he finally produced some panels for the wardrobe in our bedroom and every morning, we are compelled to spend about an hour just staring at them. There are flowers and songbirds and it's all just too wonderful for words. And the work, which is unique remember, cost less than buying off-the-peg wallpaper from Osborne & Little.

There's another advantage to the 'little man in the

village' thinking as well. Because if you have any money in a savings account, all you'll be able to watch in the coming months is the value of those savings going down and down and down as inflation bites. Eventually, you won't even have enough to buy a pack of Lurpak.

Whereas if you buy something beautiful, now, you'll be able to look at that instead. Sure, you'll be hungry and cold but you'll be surrounded by stuff that gives you pleasure, which is better than being hungry and cold and surrounded by nothing at all. Except a Dolce & Gabbana watch.

AUTUMN

Three from the sun

The water companies are having a spot of bother at the moment. It seems they are allowing all our drinking water to escape into the earth because they're too busy filling every picturesque trout stream with turds, while coating every south coast surfer with a thick veneer of human effluent.

Needless to say, the nation's beach dudes, wild swimmers and fishermen are not terribly happy about this state of affairs and are consequently rushing about saying that water company bosses should hand back their billion-pound salaries and then walk naked through the town centre while people throw vegetables at them and chant 'shame'.

I'm not sure, however, that we should be so quick to judge. As a kid I would spend most of my holidays in Swaledale, trying to dam Thwaite Beck. And what I learnt is: you can't. Water is as unruly as a railwaymen's trade union. If you plug one gap it will simply find another. And then, if it thinks you are getting somewhere, there'll be a deluge in the hills and the beck will be instantly transformed into a rampaging brown torrent full of tree trunks and boulders, and soon every scrap of evidence that you were ever there will be gone.

Water is a Terminator. It absolutely will not stop. So the idea that it can be controlled and steered and cajoled by a bunch of businessmen with a PowerPoint presentation and some meeting biscuits is laughable. They may do a pretty good job most of the time, but occasionally the aquatic Arnie will throw them a curveball, and pretty soon there will be a brown plume of excrement gushing into the oggin and a surfer will be on the news saying, 'I've got a tampon in my hair.'

I have absolutely no idea how the water companies control sewage. Or how they find any employees. Because I would rather do anything – even that – than work in a sewer, chiselling fatbergs from Bazalgette's stonework. I also have no idea why the system breaks down from time to time. But there can be no doubt that the fundamental cause is not Boris Johnson or climate change or Brexit or any other fashionable left-wing excuse: it's that this country produces approximately 70 million turds a day. Which is too many.

That's probably why, this morning, Kaleb came bouncing into the farm office to announce that he's just done a deal with Severn Trent to buy some of those turds, which he's going to use as fertiliser. He was very excited, saying it cost ten times less than animal muck and about a thousandth of what I pay for the chemicals I use at Diddly Squat.

Some quick research revealed that, according to the United Nations, 80 per cent of the world's 'wastewater' is pumped into the oceans. Which does raise a question. Instead of polluting the sea and the baby turtles that live in it, why not use our waste as fertiliser?

That brings up an interesting point. Would you buy food if you knew it had been grown using human excrement? Or, to be more accurate, how do you feel knowing that if you've bought organic veg from certain countries you've already tucked into one of Manuel's No. 2s?

Yup. People used human waste in the past, and Matt Damon did it in the future, on Mars. Meanwhile, in present-day North Korea, where the chemical fertiliser shortage is acute, they have been forcing households to produce 200kg quotas of excrement, unless one member works in a state factory, in which case they are expected to produce 500kg. That's half a ton of what they call 'night soil'.

This is then used to fertilise the fields, and that's fine, except a soldier who defected recently was found to be full of parasitic worms, several of which were more than ten inches long.

I have other issues too. Humans tend to eat plants,

which are fed by sunlight, which means they're once removed from the sun. We also eat animals that eat plants, because they're 'two from the sun'. We do not, however, eat animals that eat other animals, because that means eating 'three from the sun', which in general is a health no-no. So how do I feel about eating plants that have been fertilised using the excrement from a human meat-eater? Same as I would about eating food fertilised with dog shit. Nervous.

In the UK we do not use the term 'night soil'. It has a less sinister name: cake. And it's not raw sewage. It's treated and blended and pasteurised and possibly sold with added lavender to make it smell so lovely you won't know whether to put it in your window box or your knicker drawer.

Naturally the manufacturers are keen to talk about the benefits. Human waste is rich in potassium and nitrogen and, it's said, so much phosphorus that if it were properly collected it could account for 22 per cent of the total global demand. This makes eco-mentals jump up and down clutching themselves with excitement.

I see the appeal too because, without any mining and without any need for giant chemical plants in Russia and China, we are using an unending waste product to make more waste product.

Naturally there are some issues, because human waste

matter does contain some heavy metals that you really wouldn't want to spread on the fields, and we do take a lot of drugs and medicine. Filtering processes get rid of some of the artificial hormones used in the pill, but there's a worry that enough will slip through the net to make lady animals and fish barren. And no one wants to see a coked-up cow.

Legislation and testing are already helping minimise the risks in the UK, but even here, where rules are heavily enforced, the water companies are prone to the odd mistake, so we have to assume the same will be true of the PowerPoint-and-biscuit men who take their waste and turn it into fert.

And then there's Johnny Foreigner, who can write as many rules as he likes, just so long as they are not enforced.

It's rare that I find myself on the fence, but I do with this one. I can see the benefits. The possibility that less sewage gets spilt in the sea. And the certainty that we will have less reliance on the chem giants to produce artificial fertiliser.

But I have a genuine problem with shit. I struggle with soiled nappies and simply cannot clean up dog eggs without gagging. So the idea of eating food that I know contains human excrement fills me with revulsion.

Maybe if I thought the waste had been produced by

a 'two from the sun' vegetablist I'd feel better. But I doubt it.

Perhaps I should turn to the actress Sarah Miles for reassurance. She is now 80 years old but has skin so flaw-less she looks half that, and she's still writing and still working. And she has drunk her own urine for much of her adult life.

True heaven is a place on earth

The motor show at Earls Court in London was always the highlight of my year. My dad and I would stay at a nearby hotel in Barkston Gardens and we'd eat baby octopus at Il Palio di Siena, a small Italian restaurant where the chef would go on strike every night and the pepper grinders were the size of Apollo moon rockets.

For a small boy from Doncaster, this London living was as exotic as a St Petersburg ice palace with Julie Christie in it, but it paled in comparison to the show itself, which was a kingdom of metal, a eulogy to internal combustion and a cathedral of everything in my life that mattered.

Much was made back then of the semi-naked girls who'd disport themselves on the bonnets of the Sunbeam Tigers and Lotus Elans, and it's true they may have been a distraction for my dad, but I was far too busy collecting brochures that I could cut up to paper the walls of my bedroom. And stickers. And memories.

The man from Volvo allowed me to actually sit at the wheel of an 1800ES and, while there, I discovered for the first time that my penis was a dual-purpose organ. Then the man from Peugeot showed how you could raise and lower the windows of a 604 using electricity. Which was almost frightening. Back then we didn't have electricity in Doncaster. Or octopuses. Or pepper. It was all just coal.

I continued to love motor shows in the early days of my professional life too, travelling to Detroit every January to gawp and laugh at the nonsense Uncle Sam had created, and then to Geneva in the early spring, where you could see Sbarros and Bugattis and all sorts of stuff you were never going to encounter in real life.

Today, of course, motor shows are no longer a thing, chiefly I suspect because battery-powered cars aren't really cars at all. They're electrical appliances, like toasters. And who'd want to go to a toaster show?

Not me, that's for sure. I prefer events like the Goodwood Festival of Speed and the Silverstone Classic, where you can see real cars doing what they're supposed to do – turn petrol into the holy trinity of noise, smoke and speed.

Most of all, though, I prefer the Moreton-in-Marsh Agricultural and Horse Show, which is held every year on the first Saturday of September.

It's one of the largest one-day shows in the country and while there are cars to look at, they're manly and rugged and designed for people who have big forearms and only care about cheap parts, how many bales will fit in the back and towing limits. There are also tractors and muckspreaders and pigs and sheep with four horns and stalls selling dog treats and sensible clothing.

It's my new favourite show in the world and I wanted to see it all. But I arrived through a back flap in a large marquee that was full of goats. And I became so consumed with their adorableness that I was there for hours, talking to enthusiasts about milk and cheese and how a shoal of goats can clear an acre of brambles in less than an hour. I also discovered you can buy a billy goat for less than a tenner and realised I wanted some even more than I wanted a BMW 3.0 CSL. Then I arrived at a pen full of people and realised I was supposed to be in it, not mooching around among the animals with the best-in-breed judges.

I beat an embarrassed retreat and soon found Kaleb, who had his own stand. It was exactly like the stand Ferrari had at the Geneva Motor Show in 1988, apart from in every single detail. Furnished with straw bales, it was a tarpaulin draped over some fence posts, and it was full of men in checked shirts who were queueing up to rent his mole plough.

In the retail area, everything was made from wool, sheepskin, tweed and leather; then later, while I was enjoying a wellness pint, a Lancaster bomber flew over-head, its mighty Merlins sending a tingling up everyone's tingly bits. Suddenly, I thought I might be in a weird *Stepford Wives* version of Brexit heaven.

The whole event was as diverse as a 16th-century Ice-landic wedding and there was a very real sense that everyone had turned up with the pronoun they'd had since birth. I saw no artisanal bread or milk made from almonds and no one I spoke to said they'd 'reach out' to me later.

So, speaking as a man who's always liked the Euro-pean café culture and poncey restaurants in London and who still thinks we should be in the EU, I should have been miserable. But I wasn't. And I think you would be happier as the country slides towards ruin if you too embraced the Moreton show mentality and channelled your inner Morris Traveller.

Think about it. You're worried that your water has sewage in it, but the Moreton people weren't because they get their water from a spring. You're also worried that soon you will no longer be able to afford your early morning skinny latte, but if you become the sort of person who can make one teabag last a month, this will be of no consequence.

There's more. If you spend your life at parties full of exotic frocks and people drinking martinis, you'll be consumed by a need to keep up, but if everyone you ever see holds their trousers up with baler twine, you won't.

It's the same story with food. You've become accustomed to having options every day and you're frightened of losing that luxury. Why? I've had blackberries and marrow every night now for the past two weeks because that's what's ready in the garden. Soon I shall be eating turnips and beef and then in the spring I'll have some lamb. My variety is seasonal.

Immigration? Don't know what you're on about. Seeing a doctor? I don't need to worry because there are weeds and flowers in the woods that can cure and mend most things. Energy? Sure, the prices are scary if you rely for your warmth on gas, but Moreton people have the option of chopping logs to keep warm and then burning them when they're tired. I know, of course, this is illegal, but so is doing 80 on the motorway and we all do that from time to time.

Since I started farming three years ago, my entire outlook on life has changed. I prefer to wake up at dawn rather than go to bed then. And instead of wanting to see big things like the Hoover Dam, I now get excited by the turning of the leaves and the smell of wood smoke and a flock of yellowhammers.

Age is partly responsible, of course, but by learning

there's nothing I can do about the weather or how grain prices are affected by events on the other side of the continent, I've become more patient and calm.

And that, really, is the difference between the motor show and the Moreton show. One is full of stuff that makes you want to work and thrust and duck and dive. The other makes you realise that a mushroom you found in a dew-kissed autumnal field costs absolutely nothing at all.

A twitcher's lament

It's a target-rich environment for Britain's newspaper columnists at the moment. We've had Truss and Biden and Putin and all those vegetablists called Tarquin who spent last week emptying cartons of milk into the carpets at Fortnum & Mason. Then we had those Herberts dangling from a bridge and throwing soup at paintings and glueing themselves to the road, while the police just stood there, growing their moustaches.

Honestly, I feel like Christopher Plummer in *Battle of Britain* when he spots an unprotected squadron of German bombers trundling across the North Sea. I want to engage combat power on the mighty Merlin and radio Rod Liddle and Camilla Long to say: 'Help yourself, chaps. There's no fighter escort.'

But while deciding which of the big cumbersome bombers to go for, I noticed a flock of seagulls following a tractor across a field in the valley, and it occurred to me that this, right now, is probably the biggest story of them all. Because, sticking with my *Battle of Britain* theme, this seemingly harmless spectacle is the lone Messerschmitt hiding in the sun. The death-dealing problem that no one's noticed yet. And probably won't until it's far too late.

At present many people are thinking hard about the number of people who are coming across the English Channel every day in the sort of small boat that John Noakes used to buy for the lifeboatmen. Some are worried by why they're coming here, while others are concerned about where on earth they're all going to live when they arrive.

But there's another migration story, which is even bigger and even more distressing. Especially if you're a puffin.

We all remember what happened with Covid. Someone in China ate a bat, and then his mate got on a plane and that was that. Within weeks every country in the world was shut down, economies were crippled and millions died.

But there's another virus, which doesn't need a plane to circumnavigate the globe. It can simply travel in its host. I'm talking about bird flu. And round about now millions of geese, swans, fieldfares, short-eared owls and shovelers are arriving in the UK. To replace all the swallows, swifts and warblers, which left at the end of summer. This means that up there in the big blue it is a never-ending conveyor belt of endlessly migrating disease, misery and pestilence. And it's worse now than at any point in history.

Some ill-informed socialists say this is because of climate change. Others argue it's all the fault of rich people who shoot pheasants. And you don't care either way because bird flu isn't really an issue for humans. Unless it mutates. Or a dead bird lands on your head. In the whole world only 865 people have caught it so far, and while half did in fact die, some others had much the same lack of symptoms as with chlamydia.

Sure, you've heard that bird flu is causing chaos among the nation's poultry farmers, but, again, you're uninterested because you can eat the egg from an infected bird and you'll still be fine. Maybe you'll find it difficult to get a turkey this Christmas, but, again, that's no biggie. You'll have some beef instead, and in the meantime you'll busy yourselves worrying about the next prime minister and Putin and the eco-Herberts in hammocks.

Well, steady on, because the number of birds dying from the latest outbreak of what's technically called H5N1 will spin your head round. First identified just eight years ago, on a goose farm in – you've guessed it – China, it's now everywhere.

In Florida one sanctuary has lost 99 per cent of the birds it housed, and in the States as a whole the number of hens that have died of the disease so far stands at 41 million. And still the migration goes on.

Official figures say that this summer alone on the Isle

of Man it's killed a peregrine falcon and 200 seabirds. Twitchers there say the numbers are far higher. They talk of beaches that are carpeted in rotting corpses. Officials in Cornwall say they've collected 394 carcasses since August. Here in Oxfordshire I would usually see five or six red kites a day and maybe a couple of buzzards. But I haven't seen either for months. It's scary.

But it's in and around Scotland that the problem is at its most acute. Thousands and thousands of terns, gannets and guillemots have died. Scientists say that the great skua population is already down by 85 per cent on some islands. And because seabirds take years to find a mate and reproduce, it's feared the world may lose many entire species. Completely.

And not in a nice way. We may not have many symptoms if we catch it, but birds do. They get a swollen head, excessively watery eyes, a loss of balance (which is bad when you're 500ft off the ground), the tremors, swelling, haemorrhages and discoloured stools. It's a bad way to go.

And it saddens me. I know I'm weird, but I can spend hours watching gannets diving into the sea, and the Seychelles has been my holiday location of choice in recent years because I like to spend my days in a tree watching the almost translucent fairy terns feeding their young. Who are born wearing what look like wellington boots.

I was looking forward, as I headed into old age, to spending my days sitting in a rocking chair, watching the fieldfares and the lapwings come and go, but as far as I can tell, all that'll be left will be pigeons, which for some reason are immune. Probably because they're just rats with wings. And, to make matters worse, I probably won't even be able to afford the rocking chair.

I wanted to believe that the powers that be had a plan to make everything better. But I've just waded through a massive official report on the problem, and what it says is that protecting seabirds is a high priority for the government. Somehow, I seriously doubt that.

David Cameron murdered my tractor

My tractor has exploded. A mate who lives in the next village rang saying he needed it to mow his paddock. I won't give you his name, save to say it begins with a D. And ends with an 'avid Cameron'. But that's not important. What is important is that he borrowed it and it blew up.

I know that these days people say their computer has 'blown up' when they mean that it has frozen. Also, when people say their car has 'blown up', they don't mean it became a fireball on the A303. They just mean it juddered to a halt. Possibly because it ran out of fuel.

Cars very rarely blow up. Computers never do. But my tractor did. It wasn't the Lamborghini, you'll be surprised to hear. It was the old Massey Ferguson. My neighbour arrived to borrow it one sunny Saturday morning and set off down my drive, looking like a politician who's pretending not to understand the countryside. Polo shirt. Ear defenders. Peaky Blinders hat. And black wellingtons.

And 200 yards further on there was a bang and a

Ukraine-sized mushroom cloud. Oil splattered into all the blackberry bushes and bits of iron were to be heard landing several minutes later. I was dumbstruck because initially I figured he'd been hit by a heat-seeking missile.

Further investigation has revealed that two three-inch holes had been punched in both sides of the engine block. He claims of course that he didn't do anything wrong, that he'd simply been driving along when it decided, for no reason, to have a smoky aneurysm.

I'm actually rather sad about this because the little red Massey was built in 1961, just a year after I was born, and since then it has spent every single day bumbling about in the countryside, helping to feed and shape and strengthen the backbone of this sceptred isle. It was out there for the Cuban missile crisis and all of America's involvement in the Vietnam War. Moonshot, Woodstock, Watergate, punk rock, it saw the Beatles come and go. This honest workhorse, a simple blend of internal combustion and steel and toil, had seen us join the common market and then, while being driven by the man who precipitated our departure, it died.

However, because I bought this tractor as a Christmas present for Lisa, and I'm a sentimental old sausage, I can't just scrap it. I shall have to bring it back to life. I shall have to pay for a heart transplant and at great expense find a mechanical surgeon who can perform it.

I have similar problems with my 17-year-old Range Rover. Both its turbochargers 'blew up' recently and the cost of replacing them far outweighed the total value of the car. But sending it off to be turned into pots and pans was too much of a wrench. It would be like executing your dog because the bill for mending its broken leg would be bigger than the cost of the animal in the first place.

Of course I could afford to replace the old Range Rover with the latest model. I've tried it and it's a damn good car. But I'd have to wait 17 years for it to become 'my' car. Until then it would just be a tool.

I suspect proper farmers don't think like this. And I understand why. If you don't host quiz shows or write newspaper columns, finances out here in the sticks are extremely precarious, so the bottom line takes precedence over affairs of the heart. You can't be sentimental about an animal, and you certainly can't be sentimental about a machine. If it's costing money, get rid of it.

So on that basis I'm going to have to get rid of the Lamborghini tractor as well.

I'm well aware, when I announced, on the *Clarkson's Farm* television show, that I'd spunked £40,000 on a Lamborghini, lots of metropolitan commentators thought I was showing off. But if you know anything at all about tractors, you'll know that £40,000 for a machine that powerful was a steal. Prices for a fairly modest new machine start at twice that. An off-the-peg Fendt with the same sort of grunt as my Lambo would be about £250,000.

I know this because Kaleb keeps telling me. Every morning he arrives on the farm with a brochure or news of a deal he's heard about. He desperately wants me to get rid of the Lambo, which he says is 'shit' and too big and, in recent weeks, too 'not here'.

It has actually been fairly reliable since I bought it three years ago, but I began to notice in the summer that the brake pedal was getting a bit soft. And then it got very soft. And then it was so soft that I had to drive around with a trailer on the back and use its brakes to slow me down.

Then one day Kaleb realised that none of his tractors was evenly remotely powerful enough to lift his new disc, so he crawled round when I wasn't looking, unhitched the trailer from the Lambo, attached his kit and set off down the drive.

Imagine his surprise when he reached the bottom and couldn't stop. Once he was out of the hedge he ordered me, quite firmly, to book the tractor in at a local dealer for a service, and even though this was two weeks ago it has been there ever since. And a farm without a tractor – not even a little one because a former prime minister blew it up – is like a restaurant without a stove. It's not really viable.

And it's going to get even less viable soon because I'm told the Lambo's clutch is on its way out and replacing that means splitting the entire machine in half. I can't imagine they'll be able to do that in 20 minutes.

Every single shred of common sense, then, says I should sell it and buy something that works. But I can't. Because Diddly Squat Farm and that tractor are joined at the hip. It's like when Led Zep lost its drummer. It wasn't Led Zep any more, so they all went home.

Kaleb says that maybe I could keep the Lambo as an ornament, but the council definitely wouldn't allow me to do that. I once put a Lightning jet fighter on my lawn and tried to argue it was a leaf blower. But they weren't having any of it. So they'd definitely order me to lose the Lambo if I turned it into a fountain or something.

I shall therefore keep it going and my farm will soon be like my wardrobe. Full of shirts that no longer fit, and never will again, and jackets that are full of holes. But which I can't take to the charity shop because each one of them brings back memories of things I've done and places I've been and curry I've spilt.

The life humanity has created for itself means we are all tempted by the new and the shiny. But for me the old and the worn and the precious is always a lot more appealing.

I've become an eco-farmer – and
Kaleb hates it

In theory, this new-fangled trend for 'wilding' sounds tremendous. Because instead of spending the weekend mowing the lawn and deadheading your roses, you put nature in the driving seat and go to the pub. Where you tell everyone that you're at the cutting edge of green thinking.

I'm not sure, however, that it works terribly well as an alternative to farming. Most people imagine that if farmers would just leave their fields alone, they'd soon fill up with oak trees and wildflowers and butterflies. But I've learnt in recent years that if you leave farmland to itself it will soon be swallowed up by brambles. Except for the bits that are colonised by gorse. So instead of meadows and people in floaty frocks drinking lemonade, you end up with a badger-filled tangle of thorns, disease and the anguished cries of a thousand dying hedgehogs.

There's another problem too. If farmers stop farming and give their land back to God, where exactly will our food come from? Factories? This question has been troubling me all year. As fertiliser prices went through

the roof, past the chimney pots and the lightning con-
ductor and then continued going upwards, from £200 a
tonne to more than £1,000, I started to wonder if I'd be
financially better off if I took up wilding and went to
bed for a year.

I mean, I'll have to spend maybe £130,000 on chemicals, which might be justifiable if we got good weather
and grain prices stayed high. But what if the weather
isn't good? What if the war in Ukraine ends and grain
prices tumble as a result? These are big risks and there's
a lot at stake. You might put a £2 bet on black or red at
the roulette table. But farming this year is like putting
£10,000 on zero.

I couldn't really afford to use fertiliser, and I couldn't
afford not to because the soil round these parts, and the
altitude, conspire to make the organic option a non-
starter. Maybe I really should just say phooey and not do
any farming at all. But what if everyone did that?

I then met a local farmer called Andy Cato. He's from
the same neck of the woods as me in South Yorkshire.
He even used to go scouting in the woods at the bottom
of our garden, before he went on to form Groove
Armada. But after we emerged from memory lane he
explained that he'd had some success at mending soil,
looking after nature and growing food – all at the same
time. It's called regenerative farming and on paper it

makes no sense at all. Unless you've been in a very suc-
cessful band or you host some kind of global car show.
Or you invented the Hoover.

To help me understand, he came up to Diddly Squat
and invited me to dig a small hole in the middle of a
field. We then looked at the soil and even I could see it
was pretty feeble.

If it was a person, you'd assume it had very terrible
jaundice. Then we dug a similar hole by a hedge at the
side of the field, and the difference, even to a motoring
journalist, was obvious. Because here the soil was brown
and soily and full of worms.

My new best friend explained why. In nature lots of
different plants all grow next door to one another. It's all
as diverse as the people in a modern car commercial.
And this is what makes the soil healthy, because each
plant is contributing in some way to the good of the
whole. And yes, I know this does sound like the fore-
word in an XR recruitment pamphlet.

But since farming was mechanised farmers have only
ever grown one crop at a time. I could tell from my
mate's face that this was a bad thing, so I frowned and
nodded sagely.

He then said that what I should do is grow two crops
simultaneously. Wheat and beans. What I thought to
start with was: how do you plant two crops at the same

time? And then I thought: if by some miracle we do get the seeds in the ground, and they grow, how on earth do you turn them into food? Combine harvesters are clever. But not that clever.

And then things got even more complicated because it turns out you don't just drive around spraying the crops willy-nilly with insecticides and pesticides and fungicides. You test them regularly to see specifically what they need. And then you give them that. It's not organic farming. It's a sort of halfway house that recognises we need to eat and we need to look after the mud. And that fertiliser is £1,000 a tonne.

Kaleb was extremely unpleased about this development because driving around in his tractor, killing weeds and insects and pumping chemicals into the soil is how he makes a living. It's what he thinks of as 'traditional farming'. But in 'traditional farming' what the farmer thinks of most of all is yield. Global prices are normally set by people in suits in Chicago and he can't do anything about that. So instead he concentrates on how much he's growing. Volume is everything. The tonnage per acre.

With regenerative farming, however, you think about pounds (£) rather than pounds (lb). You accept that you grow less, but recognise that because you are using less fertiliser, insecticide and diesel, your profit margins are higher.

And to make that an even rosier prospect Andy has formed a business with George Lamb, a former TV presenter. It's called Wildfarmed and it'll pay a premium for crops grown their way. I've had meetings with them about this and it's fun because Andy is 6ft 8in, George is 6ft 6in, I'm 6ft 5in and Lisa's 6ft 2in. So when people see us they think they've accidentally arrived in the Land of the Giants. Lisa especially likes it. She says we make her feel normal.

Whatever. They swayed me. I haven't bitten the whole bullet but I did decide to try the scheme in one field. So a few weeks ago Groove came over and even though Kaleb stood on the sidelines saying helpful things like 'You've brought the wrong drill', he got the beans and the wheat planted.

And even though I suspect Kaleb is going out at night and weeing on them, I get up every morning and gaze from my bedroom window at a field that, contrary to my initial scepticism, is actually turning green. Stuff is growing.

There's an army of pigeons who are not helping, and most days the skies go black for a while as the seagulls arrive for a feast. Plus, there's the question of how well the seeds were planted in what is very stony, brashy ground. 'Well sown is half grown' is the expression, I believe. So it may yet turn out to be a disaster. But what if it isn't? What if I get a reasonable crop that has cost less to grow than usual, and that I can sell for a good price? And what if it really is helping to repair all that damage I've done to the soil over the years?

That's got to be better than the only other green option: turning our green and pleasant land into a gigantic bramble bush and then, after the badgers have eaten all the sheep and killed all the cows, starving slowly to death.

WINTER

(Lisa is) Armed and dangerous

In the world of television, you aren't allowed to have a day off work due to illness. I've seen cameramen filming a scene with diarrhoea pouring down their legs and sound recordists using their big fluffy microphone covers to deaden the sound of their explosive vomiting. I've seen producers clawing their way to work, like Alec Guinness at the end of *The Bridge on the River Kwai*, and researchers with hangovers so bad you could actually see their heads throbbing. Because we all know the rules. If you are breathing, even faintly, you go to work.

There's no fat on a film crew and no redundant parts. Everyone has a role to play, and everyone knows that if they don't show up, the big, expensive filming machine will grind to a halt. So it doesn't matter what's happened, you pick up your severed leg and you cauterise your gouting arteries and either you tip a pensioner out of his wheelchair and nick it or you start hopping.

I was once on a *Top Gear* stage in South Africa and, in front of maybe 20,000 people, I saw Richard Hammond retch. Mildly concerned about this development, I asked if he was OK. 'I've been sick in my mouth,' he replied, 'but don't worry, I've swallowed it.' That's how it must be.

When I caught pneumonia I made sure it was on the first day of my holiday so I'd miss that, not work. And I timed my first bout of Covid to coincide with the Christmas break. When I caught it the second time I manned up and just kept on going.

I assumed of course that all of this would put me in good stead for my new life as a farmer. Because you can't take a day off if you have animals to look after. No matter how terrible you feel, you absolutely have to get up and get going.

How wrong I was. Kaleb called last week to say he'd never felt so rough and that he was going to bed. And then two days later a couple of other behind-the-scenes guys who work here called to say they'd gone down as well. One even ended up in hospital. It's some kind of viral thing apparently.

And what it meant was that, for the very first time, the farm was entirely in the hands of Lisa and me. We were Ted Striker and Elaine Dickinson from *Airplane!*. The crew was down. It was now all up to us to save the day. And it was pouring with rain and blowing the sort of gale that causes wheelie bins to trigger speed cameras.

Lisa responded to this challenge by coming down to

breakfast in a ballgown and tiara. 'I can't possibly help you today,' she said. 'Look how I'm dressed.'

But she had to help me. She had to get changed and get real because the new pigs were due to arrive. And we had to make a fence to keep them here. And some houses where they could shelter from the truly biblical weather. And this is not like fixing a shelf. It's not a job you can simply put off till another day.

First of all we had to get all that we needed from the farmyard to the field we'd selected as a new home for the shandy and blacks*, a local pig breed so rare that just a few years ago there were only a handful of boars left in the entire world.

I had to drive the tractor, which was fitted with a fence-post knocker on the back and what looked like a Second World War hospital bedframe on the front. Which meant Lisa had to bring the gates and posts on our JCB telehandler.

She's never really driven it before. I'd offered to teach her, but she always arrived for the lesson in a very short miniskirt saying she couldn't possibly climb up the ladder into its cab as it would be undignified. This time, though, there was no way round the issue. She had to do it.

And I know what you're expecting. That she made a

* This may be a lady drink in the north.

total Horlicks of everything, and we had to rescue the loader that night from a river. Wrong. She was brilliant. A complete natural. It may be the Irish in her, but it was like watching a yellow hydraulic ballet as she pirouetted around the farmyard, scooping up all we'd need to make the pen. She really has missed her vocation in life, that girl. She should have been called Derek and got a job with McAlpine's. I was strangely proud.

And then, two hours later, I was even more proud because there she was, all 18 foot of her, on her back, in the mud, trying to hammer a rusted-up pin out of Kaleb's Second World War bedframe. I tried calling him to ask if there was a trick to this, but he just made gurgling noises and that was that. She had to work it out for herself, and did.

And then she emerged from the loader with a gas-powered nail gun. How best to describe this thing? Hmm. 'It's absolutely f***ing lethal' doesn't quite cover it. The nails come in a magazine format, like the bullets in a gun, and slot into place with the exact same metallic clunk you get when you cock a well-oiled rifle.

Then you gently push the weapon up against the fence post, which releases the safety mechanism and energises the gas. And then you pull the trigger and blam. People used to post videos online of girls in bikinis firing machineguns. It wasn't cerebral or erudite in any way, but it was popular. And let me tell you, Lisa with that nail gun was better. I could have sold tickets.

On and on she went, fastening the stock fencing to the post with her gun until she instructed me to drive forwards in the tractor, which would provide some tension. Sadly I drove forwards a little too far and ended up pulling the post, with the stock fencing attached, clean out of the ground. Lisa was very cross with me, and had a nail gun.

So that night, after I'd made our supper and fed the dogs and lit the fire and emptied the dishwasher, I sat down in front of a programme about estate agents in California selling hideous houses to orange people and reflected on the day's events. And what I concluded is that by succumbing to an illness Kaleb had managed to prove that Lisa can do his job just as well.

The next day, over a hearty breakfast I'd made, I broke the uncomfortable silence by suggesting to her that she might like to work on the farm more often. She responded by going upstairs for a minute and then coming down in a Chanel trouser suit. 'No,' she said.

So now it's just me. On my own. And, honestly, I'm sitting here praying that whatever is taking everyone else down comes for me too. And soon.

Bleak House

Well, this is it. The first Christmas in our new house. A house that's been built from scratch. I started designing it, in my head, about ten years ago, and then, after two architects had translated those thoughts into technical drawings, and the old house had been blown into oblivion by the largest explosion on British soil since the Second World War, the builders moved in. And now, despite Covid and Brexit and the war in Ukraine, and all the shortages all those things caused, it's finished. And I'm able to pass on a few tips in case you are thinking of doing a Grand Design yourself.

First of all, the boot room you have in mind will not be big enough. The one I built is far larger than my first flat, but it's still hopelessly inadequate. Coats and wellies and dogs and gun safes and hats and scarves and all the apparatus you need for a country walk are bulky. So that's the first thing. Make it the size of a football pitch. And even then, accept you will need to employ someone who normally designs the interiors of caravans to create a labyrinth of cunning little cubby holes where your weirdly large collection of tennis rackets can go.

This brings me on to the next thing. Everyone, when they are designing a house, has summer in mind. They visualise garden parties and Pimm's and people sitting in the shade of a honeysuckle-covered pergola, listening to bees. I did. Which is why every window is a French window.

That worked well in the August heatwave, but you need to remember that in Britain it's mostly not summer. It certainly wasn't summer last week. And that's when I discovered that some of the woodwork seemed to have warped. There's such an enormous gap in one door that you can actually go outside without opening it.

This makes security a bit of a nightmare. I realise this shouldn't be a worry, because a burglar would have to climb over the mountain of ski equipment that won't fit in the boot room before he got to anything valuable. But it is, which is why Lisa and I have to take it in turns to go to parties while the other sits at home nursing a shotgun and a heavy-ended ebony stick. This is my second tip, then: fit small doors, made from something that won't bend. Steel, for instance. Or the remains of Mrs Thatcher's spine.

The other drawback to doors that won't close properly is the cost. To try to counteract the howling gale that blows through the gaps, I'm having to write this wearing gloves.

Now. Water. We are not on the mains here, so we sank a borehole that sucks up a primordial ooze. It is possible to extract water from this in the same way as you can extract water from an iguana if you feed it through a mangle, but the water from our borehole is one part hydrogen, one part oxygen and one part dead brontosaurus.

To remove the poisonous dinosaur juice from the equation, I fitted a reverse osmosis system, and the result is: we go through a filter every seven seconds and we now have to get all our water from a spring. I spent thousands of pounds on that borehole, and now I'm drinking from a spring like a medieval peasant. Or a cow.

My next tip is the most important: do not let builders build your chimneys. You need a wizard for this, because if a chimney is built using science and angles, instead of mystery and the dark arts, it won't work. Mine really don't. I light a fire at night and within ten minutes I have a pretty good idea how Joan of Arc felt in her final moments.

Next up. Tech. Avoid it. I have dropped a laser-guided bomb from an F-15 Strike Eagle, but I cannot turn on the hot plate on my electric Aga. Nor can I work the iPad that controls the underfloor heating, or the apps on my phone that control everything else. Put simply, you don't need to be able to run a bath while you're in Milan.

My main recommendation, though, is time management. It takes years to build a house, and when you finally move in, the temptation will be to unpack two pans, put some sheets on the bed and think: 'Right. We're done.' But you're not.

You need to take weeks off work to fill the bookshelves and arrange the paintings and decide where to put the 'Jezza is a c***' ornamental plate that you were given by a mate back in the 1990s. And the AK-47. And the stuffed swan. And all the other things you've accumulated over the years. Because if you don't, they will live in cardboard boxes in the already jam-packed boot room until the day you die.

Of course, if you do get cracking, you must be prepared to put the sheets on a second bed, because the rows will be measurable on seismology machines in California. Every painting you hang will be in the wrong place. Every lamp will have the wrong shade, and 'No, you can't put that AK-47 in the hall, because people will see it.' Not with that amount of smoke coming out of the fireplace, they won't. And, oops, there goes the fire alarm again. 'Well, why did you light a bloody fire?' 'Because have you seen the price of gas?' 'I can see why someone gave you that plate.'

Here's the thing, though. As I write, it's bitterly cold outside, but as Lisa argues with herself about where the Christmas tree should go – we never thought of that – the sun is streaming through the big French windows and I'm looking at all the plumped-up blue tits on the bird feeder. So, despite the defects and the mistakes and the smoke, this is a happy house. And at Christmas what more could you want than that?

Another fine mess

I've always loved pigs, so now I'm a pig farmer. Cheerful Charlie Ireland, my land agent, says this is the stupidest idea I've had yet and wants no part of it. And Kaleb was so unpleased that he got into his pick-up truck and went to Cornwall for a week.

My logic, though, is sound. Pigs are much cheaper to buy than cows or sheeps, and unlike most other farm-yard animals they don't produce one or two babies. They hose them out like machinegun bullets. So you buy ten pigs for a few quid and three months later you have ten million. That's profit, right there. Pure, naked profit.

To make the financials look even better I decided to keep the pigs in a field that was full of potatoes that had been rendered unharvestable by the summer drought. I was going to let them rot but now, thanks to my brilliant new plan, they'd be used as pig food.

Charlie responded to this argument by rolling his eyes and going home, which meant that Lisa and I had to spend a week or two learning an all-new language: pig. You might think it's easy. You've got piglets and sows and boars, and that's it. But you're wrong. In the same way that cow and sheep farmers use words that would cause arguments in a game of Scrabble, I've learnt that in 'pig' you have weaners and gilts and that when you have a sow that isn't pregnant you describe her as 'empty'.

And then we had to get into the business of what breed to buy. The choice is endless but, in the end, we went for something called the Oxford Shandy and Black. Partly because they have comedic ears that grow over their eyes, so they literally cannot see where they're going, and partly because 'shandy and black' sounds like the sort of thing a northern girl would order in a Zante nitespot.

But mostly because it's a breed that is thought to have been created in Wychwood Forest, which I can see from my kitchen window. And because a few years ago there were only a handful of boars left in the entire world. This, then, is a breed that gives the panda hope.

We bought 15 of them and they all came in a shade of metallic bronze that I last saw in 1973 on an NSU Ro 80. But unlike the NSU they are not Wankel-powered. Oh no. These guys like the full enchilada. They even use sex as a defence mechanism.

I learnt this when I had to move the man pig from a pen containing two sows that I hoped were no longer empty, into a pen containing two sows that definitely

were. He didn't want to go, and to express his unwillingness he leapt onto the back of one of the sows and
started pumping away frantically. 'Look,' he seemed to
be saying, 'I've fallen in love. Please don't move me.'

Hmm. I could see that because she was enormous
and he wasn't, he hadn't climbed far enough up her back
to get connected. But he couldn't see this because his
ears were in the way. And I doubt he could sense it either
because his penis is so thin, there couldn't possibly be
any room up there for nerve endings.

A pig's penis is interesting. We hear often that it's
shaped like a corkscrew, but when it's aroused it's more
like a really long pipe cleaner. And, cleverly, it can be
steered remotely like the back of a San Francisco fire
truck. Seriously. The pig has control over its direction of
travel. He can go right while it goes left. And when he's
concentrating on doing that, he isn't really paying atten-
tion to someone who's telling him to 'get down'.

So how do you get a determined and lovestruck pig
off the back of his girlfriend? The solution to this tricky
problem presented itself moments later when one of
the empty sows from the other pen arrived on the scene.
'Hmm,' thought the boar, 'fresh meat. And she's more
my size.'

He immediately climbed from the back of his larger
girlfriend, and in no time at all was giving it the full Barry

White with the newbie. This made the old sow unhappy. As you'd expect. So, as the man pig unfurled his pipe cleaner and steered it into the new girl, the big pig attacked.

There then followed what might fairly be called a bitch fight. The empty sow was frothing at the mouth and frantically trying to stand still for her new man while being savaged by a much larger rival who was plainly upset that he'd broken off their engagement mid-flow and taken off with someone else. And then the second sow joined in.

I have read a few books on pig breeding, so I said to Lisa that if you massage a pig's back it thinks it's having sex and will stand still. Amazingly it worked. The attack pig I'd selected stopped trying to bite the sow that was having actual sex and stood there while I rubbed her back as she softly grunted the *Fifty Shades* grunt of pig contentment.

Lisa, however, had taken my instructions about pretending to have sex literally. She had gone round to the back of the other attack pig and was using her hips to thrust away at its rear end. This was quite a spectacle. But then, mercifully, it started to rain. And it was big, cold, sideways rain. Rain so opaque that anyone driving by wouldn't have been able to witness what looked like the scene from Dante's 17th circle of hell.

And it wasn't a brief scene either. I can tell you here that a pig is not a rabbit. It likes its sex to go on for quite a while. So the boar was pumping away, and Lisa was pumping away, and I was engaged in what I hoped wouldn't be a happy-ending massage, and we were all drenched from the downpour and covered in mud. And still the boar hadn't finished.

When he did, after about a quarter of an hour, the quantities he produced were simply unbelievable. Forget 10cc. This was measurable in gallons. And so was the splashback. I was drenched in it. And then as I stood there in horror, the pig I'd been massaging turned round and vomited explosively into my pocket. Well, you would if you'd been forced to stand there and watch your boy-friend empty his seed into the back of someone he'd only just met.

I'd been told that I should have got the empty sows pregnant with artificial insemination. Everyone said that, for a novice, this process can produce some hilari-ous results. But in my farming enterprise I've tried to keep it real and avoid stuff that I knew wouldn't work, no matter how amusing the results might be. I really didn't want to go down the Channel 5 Rebecca Loos pig mas-turbation route.

But while trying to play it straight I'd ended up as a receptacle for every sort of pig juice you can think of.

And Lisa felt like she'd been in an episode of *Black Mirror*. Plus, her new Canada Goose coat is ruined.

And one day the pig-owning experience will get worse because I'll have to send them 'down the road' to be turned into sausages and ham and pork chops and bacon. And all so you can go to the supermarket and moan about how expensive meat is these days.

Now I am become the Destroyer of Worlds

By any standards I'm a busy man. I make two year-round television shows, run a farm and every week I write three newspaper columns. On top of all this I have a brewery, a book deal and I'm in a constant battle with the local council to keep my farm shop open.

And yet, despite all this, I looked at my diary last night to see what the week had in store and it was page after electronic page of virgin whiteness stretching all the way to the end of next weekend. This was tremendous because it meant I could spend the entire time in my woods, playing God with what's called a mulcher. Although I've come to know it as 'The Destroyer of Worlds'.

Officially it's called a Robocut2 and when it arrived I was a bit underwhelmed. It has tracks, like a tank, and a snazzy yellow paint job, but it's not much bigger than a kitchen table and has a no-horsepower diesel engine that gives it the same top speed as an earwig. Plus it's radio controlled, so there are complex electronics, which is exactly what you don't want in a big wet wood that's festooned with hidden badger setts and jaggedy tree roots.

I figured that I'd actually rented something more suited to gardening than forestry. But never mind. I only needed it to clear a small patch of brambles by the pond and I reckoned it could do that. Because on the front there's a rotating drum. It's not a big drum, maybe 9in in diameter, but it's studded with steel spiky bits that are designed to remove weeds.

I reckoned I'd have the job done in a day but I was wrong. Because two minutes after slipping into some safety goggles and starting the engine the brambles were gone. And I don't mean 'uprooted'. I mean gone. They had ceased to exist. They had become dust. And I was standing there with my arms aloft like Russell Crowe in *Gladiator*. 'Are you not entertained?' I bellowed to a nosey woman who'd turned up from the village to see what I was doing now.

I was staggered. I've never encountered a machine that does quite so efficiently what it sets out to do. In a whole day I could use it to set nature back about four million years.

As I had it for a whole week, my next idea was to tackle the rides that used to run through my woods but which, thanks to the natural world's endless greed for real estate, are now impassable.

Rides are important. They are straight paths, about ten yards wide, that cut through many of the nation's well-managed woods. Installed on an east–west axis to maximise the amount of sunlight reaching the ground, they were, I assume, invented by those of a horse-riding disposition. They're probably quite good at halting the spread of a fire as well. But to eco people like me they have an even bigger role. The sunlight encourages all the diverse and sustainable things that sound good in acceptance speeches.

Doubtless a government grant would be available for the kind of ecological work I was planning. But if you apply for a grant you have to fill in 2,000 forms and wait for 2,000 years for a man to come in a rented Vauxhall and tell you that you must stop what you're doing because he has found a bat or some moss.

I therefore decided to fund the project myself and immediately plunged into the operation feeling a bit like Sigourney Weaver must have felt on the spaceship when she became a human forklift truck. I was no longer going to be cowed by nature's capriciousness. Me and Robocut2, we were going to be its master.

The ride in question runs from my local village all the way to the main road about a mile and a half away. This, then, was a big job, by far the biggest I've ever attempted on the farm.

Being careful to stand behind the machine – it can spit thorns out of the front at the speed of sound. And the sides. And sometimes the back as well, hence the goggles – I set off and soon encountered a tree. It was an elder, or a crack willow, or something like that, but whatever it was it didn't look very important and it was in the way, so I decided to see if my machine could knock it over.

It looked like a tall order for the no-horsepower engine, especially as it was using tracks for traction. Tracks don't work. I've never seen any tracked vehicle go for more than 200 yards without getting stuck or breaking down. It's a worry for the Ukrainians right now. And it was a worry for me too as I nudged up against the mysterious tree, which buckled, fell over and was immediately sucked into the drum, where, in a glorious moment of what sounded like mechanical indigestion, it was converted into atoms.

I was then carpet-bombed by a gigantic flock of wild birds. Despite the noise and the devastation, they were fluttering around everywhere. I can only assume the trees I was turning into molecules had been full of insects, which, after I'd passed by, were laid out on the forest floor like a giant smorgasbord.

And then I encountered one of those thorn-type plants that entangles itself into a tree and is quite impossible to remove. Not for the Robocut it isn't. It simply grabs the stem and yanks the whole thing out of the branches before turning it into absolutely nothing at all.

After an hour I had cleared 100 yards or more. But I was getting cocky and tackling bigger and bigger trees and even the stumps they left behind after they'd been nuked. Which is probably why the machine started to emit some sparks. I dismissed this as a bit of wood that was stuck in the machinery and which was being converted into fire by friction. But then there was some smoke. Quite a lot of smoke. Followed by the sort of noise that tells a driver his big end has just punched a hole through the crank case.

It turned out the drum had become detached at one end. But as it was rented, this was no problem. They came out, changed the head unit and the next day I was back at it. Until I ran over a really big log, a proper Robert Plant, that jammed the drum. Normally this is the sort of thing that causes me to give up, but I was making good progress and it was good, proper work, and I was good at it, so I toiled till nightfall to free it. Which I did. With the winning combination of a hammer and a blowtorch.

I could barely sleep last night knowing that once this was written I could get back out there, smashing down trees and ripping up brambles in an orgy of ecologically driven diesel and destruction. But this morning, literally ten minutes ago, I learnt that my week is no longer free because my No. 1 daughter called to say she's on her way to the hospital in Chelsea. To have a baby. My first grandchild.

Walking the dog

The greatest gift God bestowed on any animal was the dog's unwillingness to take itself for a walk. It wants to go for a walk. It loves going for a walk. Nothing makes it happier. But it will not go on its own. You have to be there too, even if it's cold and raining and you'd rather sit by the fire watching football.

And I am deeply thankful for that because when you are forced by a wave of gentle whimpering and cute, tilted-headed pleading to get off your arse and go into the woods, you are rewarded in spades. Not just because you've brought such obvious joy to your dog but because if you go for a daily walk on the same route, you start to understand that nature, left to its own devices, is cruel and stupid.

I realise of course that wilding is all the rage at the moment. There's a very excellent book on the subject by Isabella Tree, who decided to stop farming her estate in Sussex and hand it back to nature. She paints a rosy picture of turtle doves cooing in the trees and wild animals frolicking in the wildflower meadows. It makes the reader feel inclined to grow a beard and boil some wood to make a hoe.

But on my farm things are a little different. There are large areas I've left alone, expecting that they would fill up with wolves and beavers, but instead an impenetrable forest of brambles has emerged, making them no-go zones for anything that has skin. Nothing can live in them, not even bacteria.

Then there are the streams. Back in Victorian times when man sat at the head of the nature table, someone decided, quite rightly, that these water courses should be managed and controlled so they could be used as drinking pools for the sheeps. Nature, of course, didn't like this, which is why, over the past century, it has worn the dams away. To make sure they can never be mended, it then festooned the banks of the ponds with plants made out of high-tensile steel that grow out of the ground, move along for three feet and then attach themselves to the ground again. This means there's an invisible network of 'hoops' that will trip you up and plunge you face first into what's basically a cocktail-stick tree.

Animals? Well, we have grey squirrels that kill trees. Roe deer that kill trees. Fallow deer that kill trees and muntjacs that kill trees. And badgers that burrow under trees, so they fall over and die. We also have diseases that have arrived in imported trees that kill trees. If you only ever go for a walk on Boxing Day, you don't notice any of this. You have to go every day so you can properly witness the relentless destruction. Until summer, when you can't go at all because the woods are full of nettles taller than the Eiffel Tower.

Walking through the woods, then, is a constant reminder that man is brilliant and good, and nature is stupid and vindictive. And then I'm always brought round from this silent reverie by my labradors.

I don't know about England, but I'm told you are breaking the law in Scotland if you don't have your dog under control 'in a public place, or in a private space'. Private? Yup, it seems even when you are at home, you can be fined if your dog won't sit or go on its bed when asked.

Mine are pretty well behaved, but when they spot a deer they lose all sense of reason and hare off in pursuit. They've never caught one, and given a deer's ability to jump over the sort of wall Donald Trump could only dream about, they never will. But it does worry me that they can be so consumed by bloodlust.

And that's why, on this morning's walk, I was frightened half to death by the sound of an animal howling in very obvious distress. Fearful that one of my dogs was being attacked by a disease-ridden badger, I plunged into the undergrowth, running madly to the source of the sound. But I never got there. Because after just 25 metres, I was completely entangled by brambles. Blood was spouting from arterial wounds. My hat was ripped from my head by a thorn. My trousers were torn to shreds. And that's when the dog loped past, happy as Larry.

God knows what it had been doing, and thanks to the wilding (laziness) I've been deploying in that part of the farm, we'll never find out. But anyway, I'm sitting here now, wearing every Elastoplast I could find in our medicine drawer, and not daring to take off my jumper as I know it's festooned with thorns that will remove even more skin should I pull it over my head.

So let's forget walking and get round to the meat of today's missive: a review of the new 5m long, eight-seat Land Rover Defender 130. It would be a nine-seater if you could specify the front-row jump seat that's available on other, shorter models, but this would then make it a minibus and you'd have to pass a test in weirdness* before being allowed to drive it.

The extra length, all added at the back, doesn't do much for the styling – it's a very odd- looking car – but there really is space in there for eight full-sized adults.

* I'm not saying all minibus drivers are weird. Just most of them.

Providing of course those ordered to sit in the stern are mechanically minded enough to fold the middle seats down. I'm afraid I gave up.

The other drawback is when the seats are up, the boot's not very big. You certainly won't get a couple of labradors in there. I know. I tried. This means the 130, like pretty well all the 50 or so models in the Defender range, isn't really going to be much use as a practical workhorse. You need to think of it as a rival to the increasingly excellent Volvo XC90. Except for the interior trim. The Volvo feels Swedishly cool and stylish, whereas the Land Rover feels dark and foreboding. And the trim is nowhere near as tactile. Let's be kind and say they were going for a functional look.

That said, the Land Rover drives well. The diagonal pitching I experienced in the short-wheelbase 90 model was not in evidence at all. Maybe the air suspension you get on this car is responsible but whatever, it was all smooth and serene.

The petrol engine may have helped too. Of course, you don't buy a car like this with a petrol engine, even if there is some kind of mild hybrid assistance, because it feels like conspicuous and unnecessary consumption. As it's not a sports car and you're never going to drive it like your hair's on fire, you may as well go for the diesel. And if anyone accuses you of using the fuel of Satan, tell them it's a sort of hybrid and then they'll have to shut up. Or tell them you voted Labour. Same thing.

By now you probably think it's a pretty good car and you'd be right. It is. It's not even on nodding terms with the concept of pretty and the boot's small when all the seats are in place, but none of this really matters because prices for an entry-level model start at £74,000. And that's about what I have to spend to take back control of my woods from nature.

SPRING

Environmental doughnuts

Just before Britain freed itself from the EU, Boris Johnson appeared at a cattle market and, with much gusto and great enthusiasm, told farmers they would have nothing to worry about. He would look after us. We could go home and relax in a hot bath. And then the square root of bugger all happened for quite a long time.

I used to get £83,298 a year from the EU, which of course is a huge amount of money. But there was a quid pro quo. I had to sell my meat and bread and barley at a loss. It was idiotic socialism but, selfishly, I quite liked it. And so did you because it meant you could afford to feed your children, rather than eating them.

Then came news that in 2021 I'd be getting only £73,138. Then, in 2023, this would drop to £48,149 and it would keep on dropping until, in 2028, I'd get absolutely nothing at all. I'd be in a free market, which is fine, but I'd be competing with foreign farmers who would still be getting a leg up from their governments. And that isn't fine at all.

Sure, you'd still get your cheap food from the abroad,

but British farmers would be in the poor house, sucking on clumps of moss to stay alive.

And then, in 2018, two years after we learnt the grants would be going, the government climbed onto its soapbox and announced that instead of being paid money for selling stuff at a loss, farmers would get 'public money for public goods'.

This sounded great, and all over the land sixth-form beardies and ramblers jumped up and down with glee. But hang on. What are 'public goods', exactly? More footpaths? Because farmers can't sell those. And nor would we get very far if we turned up at the market with some peat bogs and a 400-acre wetland habitat full of rare grasses and wading birds. Is there a market for avocet? I think probably not because I bet it's disgusting.

It caused me to think that the Packhamites and the Mayists had wormed their way into the government think tanks, and that in future I'd get money only for opening up visitor centres to Labourite mad people who wanted to experience the healing power of bees.

Food? Nah. In a world where that comes to your door on the back of a moped, and meat is murder, it wasn't important. In the new order worms mattered more, and voles and earwigs.

As a result of all this I've been farming for the past

four years with absolutely no idea where things were headed. I didn't know what these 'public goods' were. No one did. There was just this gnawing, all-pervading despair, a sense that our future was in the hands of militant eco-lunatics who hated cows and a government that, let's be honest, hasn't exactly covered itself in glory on any front so far.

They kept saying the detail was coming but I was expecting another HS2. Another Brexit. Another broken-down aircraft carrier. But then the detail arrived in a document that's more than an inch thick. And you know what? I have a horrible feeling that I may have been right. A horrible feeling that was cemented into reality this morning when Cheerful Charlie, my land agent, boinged into my office wearing a grin as wide as Dartmouth naval college.

I can understand why because instead of one basic payment there are now 250 schemes I can sign up to, each of which will come with a raft of forms and code numbers. His billable hours will skyrocket. And I'll have to sit there with matchsticks in my eyes as he works out, to the nearest square centimetre, how much echium we are growing. And how many inches of hedge have been trimmed and how many boggy and sad places have been created. Because this is what Johnny Government is now paying for.

There's other stuff too. If I'm on the Isles of Scilly I can get £279 a hectare for grazing cattle, but I've just looked out of the window and I'm not on the Isles of Scilly so that doesn't work for me. I can get £10.38 if I make a breeding area for skylarks but that won't work either because the badgers I have on the farm will eat whatever eggs they lay. I can also get £90 for shooting deer, £50 for squirrels and £5,500 for wasting a rhododendron bush. But there is some good news: I can now get a grant to maintain the 40 miles of drystone walls that crisscross the farm.

This, however, is complicated. If I get Gerald to maintain my walls, I get £15 per 100 metres. But if he has to restore them, I can get £31.91 per metre. And what's the difference between maintaining and restoring? I don't know, but I bet the government has set up a 'wall police' that will tour the land in their rented Vauxhalls making sure I don't get the big cash for restoring when all I'm doing is maintaining.

So what about food, which, after all, is the whole point? Well, I've waded through most of the document but so far there's barely any mention of it. It's like they've said 'food damages the environment so let's not bother with it. Better the soil is used to store carbon.' Which means there will be a soil police too. And a woodland police. And a footpath police. And a curlew police. And you, the taxpayer, will have to pay for all of them, and what do I get? About half of what I got when we were in the EU.

And why is any of this necessary? When I walk round Diddly Squat Farm each day I am always overwhelmed by a sense that it's not really mine. A house comes and eventually goes. So does a car and a tree and so do we. Even the sun will one day go out, but when it does and all life on Earth ends, these high, brashy hills in the Cotswolds will still be here.

Many of us get a bit giddy when we sit in a desert and contemplate the immensity of the night sky. Well, I sometimes get that same giddiness when I'm walking around Diddly Squat. 'Forever' is a concept we humans struggle to comprehend.

And there's more. Only about 7.5 per cent of the Earth's surface is covered in soil good enough to grow the food the world needs. And I'm conscious every time I drive out of a field and leave great big clods on the road that it will be washed into the drains and then into the rivers and then out to sea, where it will vanish. Farming has done that to my head. Which is why, before leaving a field, I always find a bit of hard core or grass and do doughnuts to clean the tyres. It's environmental best practice.

In short, I feel compelled as a farmer to try to keep

this tiny bit of land in the best possible shape. I want you to enjoy the food I grow here. I want you to enjoy the views. And I want to take care of everything. And I've met a lot of farmers now and all of them think the same way.

A thing that seems very thingish

I have never really seen the appeal of paying a premium for something just because it was once owned by a famous person. I mean, last year somebody paid £650,000 for a 37-year-old Ford Escort simply because Diana, Princess of Wales, once tooled around Knightsbridge in it. In my head, that's madness. It's nearly as mad as the man who once tried to sell me a Ford Escort at a premium because it had once been owned by me.

And then you have that tech nerd who splurged $2 million on a guitar just because Jimi Hendrix had played it at Woodstock, and people who pay a similar-sized fortune for a set of golf bats used by Tiger Woods or a handkerchief into which Diego Maradona once blew his nose in 1996. Although, actually, the contents of Maradona's nose back then was probably worth a fair bit.

Part of the problem is that to make the price seem worthwhile, you have to tell people why your new acquisition was so expensive – 'It was Richard Madeley's cricket bat, you know' – and doing that will mark you out as a bit of a berk.

This brings me on to Jeffrey Archer. Sitting on a plinth in his agreeable London flat there's an old stopwatch. It's a lovely thing, for sure. But alongside it there's a folded placement card that says that this was the very stopwatch used on 6 May 1954 at a meeting of Oxford University's athletics society on the Iffley Road track when – and finally we get to the point – Roger Bannister broke the four-minute mile.

I have to be honest. When I read this, a bit of sick came into my mouth. Because why would Jeffrey want to explain the provenance of his stopwatch in writing? To remind himself every morning why it was there? Or to show off to visitors?

All that being said, my eye was caught last week by a story that suggested the original Poohsticks bridge might soon be offered for sale.

I cannot even begin to find the words for how much I adore those Winnie-the-Pooh stories. *The House at Pooh Corner* is, without doubt and by a huge margin, my favourite book of all time. I know as a newspaper columnist I'm supposed to say it's something grand like *Ulysses* or *Paradise Lost* or *Cannery Row*, but it just isn't. Because none of these is anything like as good at defining and bookending the concept of what it is to be a human being.

You can think of anyone you've ever met and they

will be pretty much identical to one of the nine Pooh characters. Whenever I'm filming *The Grand Tour*, I always think of James May as Eeyore, Richard Hammond as Piglet and myself as Tigger. On *Clarkson's Farm*, you have Charlie Ireland, who's Wol, Lisa, who's Kanga, and Kaleb, who's Rip out of the television series *Yellowstone*. He's the exception that proves the rule, I guess. Although he is a bit like Rabbit as well.

And there's more. Because these are clearly children's stories, and yet, also, they're not. My kids used to love me reading the books to them at bedtime – but nowhere near as much as I did. The chapter in which Eeyore has a birthday and gets two presents is so fantastic and so beautifully told and so unbelievably funny that I never once got to the end of it before falling off the bed in a fit of spasming laughter. I've just reread it now and I'm going to need a moment to compose myself.

A moment later.

And back to the Poohsticks bridge. It was bought at auction in 2021 by William Sackville, the 11th Earl De La Warr, who wanted to save it for the nation. Later, however, he said he'd been a 'bit stupid' because he forgot about the VAT and the buyer's premium. Which meant the final bill was £131,000.

Now he's hoping to get that back, and more, by selling it. And even though I have an aversion to this kind of thing, I do admit that I'd feel very warm and fuzzy if I could buy the actual bridge where AA Milne stood with his son, Christopher Robin, in Hundred Acre Wood and invented a game I still play and enjoy today.

I even have a stream, in a wood, where it could go. I was tempted. So I'd fully understand why someone might take the plunge. I wouldn't even mind if they carved a little Jeffrey Archer-style message into an upright, telling visitors that this was the very bridge where, on 11 April 1926, etc., etc., etc.

But there's a question mark on that, because is it? When they sold it in 2021, the auctioneers said the bridge was one of many that had been erected over a stream in Ashdown Forest. And that it had been built in 1979. Which, even to a mathematical Luddite like me, seems to be about 50 years out.

No matter. This was endorsed by Christopher Robin Milne as the official Poohsticks bridge where he and his dad used to race twigs. Which may have been why AA wrote about it. Or maybe he wrote about it first and they played it afterwards. Christopher could never remember which, and, as he died 27 years ago, we'll never know.

What we do know is that the newly restored bridge is now on a farm in East Sussex, having been taken there from the house nearby where the Rolling Stones guitarist Brian Jones sank to the bottom of a swimming pool because his liver was so heavy and drowned. So now, all of a sudden, the message you'd need to carve to explain what you'd bought is getting awfully wordy. And vague. 'This is a replica of what may be the actual bridge where . . .'

I fear that, when all is said and done, whoever buys it will be paying £132,000 for a collection of timber and some nails. And if that's all you want, you could almost certainly get a better deal at Jewson.

A total pig's ear of it

I realise of course that it would be very Jeremyish to say that I enjoy taking my farm animals to the abattoir. But like many farmers I don't. I absolutely hate it. I can never sleep properly the night before they go, and all the way to the slaughterhouse I have what feels like a hot cricket ball in the pit of my stomach. And then when it's finally time to say goodbye, I always become a little bit unmanly.

It was bad enough with the sheep and worse with the cows. But last week I had to take seven of my boy pigs to be killed and that was gut-wrenching. I know that I'm trying to be a farmer and that this is what farmers do. And I know I will enjoy the bacon and ham and pork chops that result. But it's not easy, taking seven happy, healthy pigs from their woodland home to their deaths. Especially as pig prices are currently so low I'll almost certainly end up making a loss on the transaction.

The only good news is that I was not faced with the option of killing them using the halal method. With cows and sheeps you can choose to have their throats slashed and blood drained – not that I ever would with

an animal raised at Diddly Squat – but pigs do not feature in the Muslim diet, so instead of using prayer to
make the process humane, the slaughterman uses carbon
dioxide instead. It was so quick that by the time I'd com-
pleted the paperwork all seven were gone.

There was some good cheer, though, because as I
drove home from the abattoir, I realised it had been
three months, three weeks and three days since a rented
man pig had sprayed his harem with gallons of his
sperm – incredibly, 250ml comes out on each go – which
meant that, back at base, four of my pigs should be
coughing new pig life into the woods at any moment.

Indeed, it was that night at 11 o'clock that I got the
alarm call. Pig One was nesting in her pigloo, and then
she was on her side, and then with barely a murmur out
came what was possibly the most adorable little creature
you've ever seen. It even had its own slithery membrane,
like the sausage it will one day become.

And then nothing happened. Pigs have up to 16 nipples, which means they are capable of producing up to
16 piglets. But after two hours we had just one. This
meant that I'd have to put my arm in there to see if I
could maybe help. But Lisa then said – and I promise
you she really did – that my hand was too big and that
she should do it. So in she went, glove-free, so far past
her elbow that I moved round to the pig's face to see if

her fingers were coming out of its mouth. The report was gloomy. Despite a good old rummage, she could find nothing in there.

So I called the 24-hour emergency helpline and spoke to the vet, who said I should take the mother pig for a walk.

Really? I can't imagine saying to a woman in labour that things might be speeded up if she went for a stroll round the hospital garden, but I'm a man and can only take guidance on these matters. So I got her up and, bugger me, it worked.

By four in the morning we had ten piglets, which meant that on the day I was three pigs up. And there were still three more pregnant pigs to go. So we came home and, despite the hour, celebrated hard with a bottle or two of wine.

The next night we had just got into bed when the alarm went off again. Pig Two was nesting. So off we went to the woods for another night of pig fisting and worry. This time, though, three came out in short order and that was it. Three. It was so pathetic we've named the pig Swizz.

And we've now named the first mother Clumsy because on the first night she sat on two of her ten piglets, killing them. And since then she has squashed two more. This is extremely irritating, and sad too. And expensive.

Still, at least it's better than what happened when the first pig gave birth a few months ago. She was so embarrassed about squishing her new baby, she ate it to try to hide the evidence. She probably thought she'd been quite clever but she wasn't clever enough to know that the little black box in the corner of her pigloo was a camera . . .

The final two pigs were both sows, which meant they were experienced mothers. They would give birth easily and not kill any of their children, which meant that finally we could let them get on with it as we enjoyed a full night's sleep.

Nearly right. One gave birth to ten and then stood on one of them, killing it, and the other seems to be ill. She's off her food and her bosoms don't appear to have filled with milk. God knows what's going on there. All I know is that so far 28 pigs have been born and eight have been killed. That's more than a quarter and that's one hell of a rate of attrition.

And there's another issue, because what do you do with the bodies? I don't understand refuse collection at the best of times, but I'm fairly sure the much-feared binmanieri would not take away eight flattened piglets. So I can't just put them in the bin.

My dogs looked hopeful, but I didn't want them getting a taste for my new income stream in case they decided at some point in the future to eat them before they had been trodden on. And nor, by law, am I allowed to bury them in case their juices get into the water table. Nor can I simply leave the carcasses in the fields for the red kites and the buzzards. Not that I could, even if I was allowed: since the last wave of bird flu all the birds of prey round here have gone.

I therefore turned to the government's pig police for guidance and after a waffly bit about how I'd need paperwork to move the bodies – arrest me – it said I could take them to a maggot farm. Yes, well, that's not happening. Or I could give them to the local hunt.

I chose this option as I quite liked the circularity. I raise pigs and the casualties are used to feed the animals that are then used to keep fox numbers down on the farm. That's the countryside in balance. Except, of course, since Tony Blair banned hunting, the countryside is not in balance. Which is why the foxes are now circling my pig pens with their knives and forks.

I don't know what to do about that. Actually, I do and it's very Jeremyish indeed. But we live in strange times, so I'm not going to say what it is.

Lady garden

There's a lot of jumping involved in farming. You spend a great deal of time hurling yourself off gates and trailers and straw bales, and that's fine when you are 14. But like almost every other farmer in Britain these days, I'm in my early sixties, and so are my knees. Which means that while I can get on top of things, I can't jump off them any more for fear that my legs will bend the wrong way and that'll be that for six months.

I have a similar problem when I drop something. In the past I'd simply bend down and pick it up, but my back's no longer really up to that. So now when I drop something, I have to spend a great deal of time trying to work out whether it's worth picking it up or whether it would be easier to drive to the shops and replace it.

And on top of the physical issues, which will only get worse, there are financial problems too. And they're going to get worse as well. Because the grants and subsidies that I used to get from the EU, to recompense me for selling food at a loss, are dwindling until, in three years' time, they will dry up completely.

These, then, are troubling times, because what am I to do? Farming hurts my back and my knees, and if I attempt to use my land to grow food, I'll lose money. It has been causing me some sleepless nights, that's for sure.

But I was cheered recently when I read about a former *Dragons' Den* investor called Rachel Elnaugh who'd bought a slab of land in the Peak District and turned it into some kind of peace-and-love nature reserve where people can come to get in touch with their chakras. Clever, I thought, because this is the sort of thing for which you can get grants and subsidies. Food production is an eco no-no. It causes all sorts of problems with the upper atmosphere, apparently. But inviting people with pink hair to sit in a wood and make daisy chains? The government loves that stuff because it doesn't mess with its net-zero targets.

So, intrigued by the dragon lady's ideas, I did more investigating, and it seems her operation is based in a wood. She refers to it as 'an incredible vagina of land' and says it'll make the perfect sanctuary for antivaxers. Remember them? She does for sure. She says the whole pandemic was a 'great bio-weapon' developed in Switzerland, and that the chief medical officer, Chris Whitty, 'will hang' for recommending that children have the vaccine.

And her investors obviously think along the same

lines because they coughed up a million quid, which was used to build a car park and a teepee. There are also other things there, to do with the Earth's magnetic field and magic mushrooms, which she says can open people's third eye.

This, then, is the sort of place where you go to live 'in the moment'. That's not something I've ever been able to do, because if you are living in the moment, you're not thinking about what you want for supper that night, which means that when you've finished being present, you'll find that your fridge is empty.

Whatever, I loved the idea that she'd obviously hit on a brilliant way of relieving the nation's Keiths and Candice-Maries of their money and I was thinking of maybe doing something similar at Diddly Squat. I could fill the woods with lots of big women and beardy men who could make pentagons out of twigs and dance naked round the redwoods.

But guess what? Further research reveals that poor old Ms Elnaugh has fallen foul of the Peak District National Park Authority, which says that she doesn't have planning permission to develop her 'vagina of land' and that it isn't interested in her 'shamanic wisdom'. So now the poor woman is being forced back to the drawing board. Maybe she will now start a place where people can come to do goat yoga. That's something Lisa wants to do at our farm.

She's also keen to invite paying guests to lie on a beehive so they can experience the healing powers of the buzzing noises. This is probably where farming's headed.

Or is it? Because if the future is going to be all about treading lightly and singing 'Kumbaya', then why do anything at all? That is certainly what seems to be going on in the world of gardening at the moment.

Writing in *The Times* recently, the always-astute TV producer Richard Wilson noticed that the homemade videos sent to *Gardeners' World* by viewers every week usually show gardens that have been the subject of no gardening at all. 'Look at what we haven't done,' the enthusiastic contributor says.

It's all gardening for your mental health and climate change and vegetarianism, all of which means doing no gardening at all.

And we've seen more evidence of this trend at the Chelsea Flower Show in recent years. Gone are the days when people created dazzling blooms and water features and glorious, technicolour rockeries. Now the exhibitors turn up with gardens made from things they found in a skip nestling in a bed of weeds. I used to love Chelsea but these days it's like walking through my local tip.

Perhaps this is what I should be doing on the farm. Growing moss on all the old fridge-freezers that people lob over the hedges every Friday night and making visitor centres for the pollinator community out of old tyres. Last weekend someone left a car bonnet in one of my fields – Nissan Micra in case you're interested – and what I did was: hitch up the trailer to my tractor and bring it back to the farmyard. What I should have done was: leave it there until it was buried in a forest of thistles. That would have been kinder to my knees and back, that's for sure. And I probably could have got a grant.

We all need to get a grip. Yes, it would be easy for me to let the brambles and the badgers take over my farmland and to sit back and watch the deer and the squirrels eat all the trees in my woods. And yes, it would be simple to open this thorny place to hippies and witches so that they can make druid noises and boil wood to make hoes but I do feel that I have some responsibility to put food on your table.

It hurts my knees and the outgoings this year are truly terrifying, even for me with four other income streams. So I can't hand it back to nature and I daren't move forwards. Which is why, this morning after I'd cleared away two dead lambs, one dead goat and another batch of dead piglets, I decided to plant my game covers with mustard. It's my last roll of the dice. My last chance to make something – anything – work. And if it doesn't? I don't even want to think about it.

I can't explain

I have some experience of not getting planning permission, and what I've come to understand is this: whether you want to build a conservatory, or a funeral home, or a nuclear power station, you've got to get the language right. Sustainable. That's an important word. Your conservatory may feature window frames made from depleted uranium, but that doesn't matter if you describe it as sustainable. And mental health. That's critical. You need a sustainable sun room full of eco-plants because it's good for your mental health. Plus you will empower the local building trade in a way that will be 'transformative' to the low-income 'community'.

Sadly, however, no matter how well versed you may be in modern government-speak, you will come up against a neighbour in red trousers who knows the even more powerful language of nimbyism. And he's going to say that your new conservatory will cause more 'pollution', 'traffic' and 'noise'. That's the holy trinity for those who worship at the altar of Laura Ashley. And if that isn't working, they'll wheel out the trump card: dark skies. They'll argue that your new conservatory will cause light pollution, and then, I'm afraid, you've had it. Especially if there's even a suggestion that you might harm a bat.

All of which brings me on to the Duke of Beaufort. He recently applied for permission to stage two summer concerts in the agreeable grounds of Badminton House – the Who and Rod Stewart, in case you're interested. And I'm sure his representatives used all the right words.

They'll have glossed over the fact that it's bloody expensive to run a big house and new income streams are necessary, because that sort of argument doesn't sit well in a country where anyone with a big house is wrong. That's the law. So the duke's advisers will have relegated the business angle to page 12 of the application and concentrated instead on how the sustainable, low-impact, green events will empower the low-income rural community and boost the mental health of the region's bats.

Sadly, though, the duke's neighbours are not just well versed in the language of nimbyism. They are fluent – they are past masters – in the art of objecting. So they started by pointing out there'd be increased traffic in the area and that noise would 'reverberate' in nearby villages – presumably causing many bat deaths and 'mental health issues'.

Naturally, they also said the concertgoers would engage in 'rowdy behaviour', even though it's the Who and Rod Stewart we're talking about. Most of the audience will be in their sixties, and when Roger Daltrey sings, 'The kids are all right', they'll turn to one another and say, 'They really are. Henry's a commodity broker now, and Harriet is doing ever so well at Freuds.' Then, when it's all over, they'll go back to Stanton St Quintin in their Teslas, and Keith Moon will not head over to the local hostelries to blow up the lavatories because he died 45 years ago.

Fearing perhaps the council might cotton on to the fact the audience are extremely unlikely to drive their cars into the nearest swimming pool, the red-trouser people decided then to open up with sustained machine-gun fire. Crime. Disorder. Public nuisance. Emergency services. Road safety. Pandora's box. This was the Middle England playbook, and if they'd stuck to it, they might have got somewhere.

But they got high on their own supply and became silly, saying, 'With 11 to 12 hours' drinking licences, drunks will camp overnight . . . increasing the potential for a major fire incident.'

Right. I see. So this 65-year-old reveller overdoes it on the noon balloons and the Whispering Angel, puts up a tent he's somehow smuggled into the venue and then, using some of the kindling he's brought from the wicker basket in his snug, gets a fire going, which, despite the constant rain that goes hand in hand with British summertime concerts, somehow turns into a major Australia-style inferno that completely engulfs three neighbouring villages and ruins the dark skies for miles.

It's the most preposterous argument I've ever heard. There was, once, a fire at an outdoor gig. It was caused by a faulty light on the stage and was quickly extinguished using stamping and a blanket. No one was injured and Bruno Mars was back at the mike eight minutes later. So the fire argument doesn't wash.

And I'm delighted to say the duke's local authority saw it for the nonsense it was and gave the gigs the go-ahead. And before you write in saying, 'How would you like it if your neighbour invited the Who to perform in his garden?', I'd say: 'I'd like it a lot. Especially if they bring some lasers and do "Baba O'Riley".'

I fear, however, that this is not the end of the story, because now 'sustainable' has been balanced out by 'traffic', and 'empowering' by 'light pollution', the red-trouser brigade is going to become increasingly desperate in its constant battle to keep Britain as it was in 1957.

Mr Sunak announced recently that planners will be encouraged to look favourably on rural schemes, but they're going to be up against a tub-thumping army that will quickly recognise that the fire argument was a bit of an oxbow lake and will start to argue that the new housing estate for the low-income community will cause a plague of luminous locusts that will spoil the dark sky. Or that it will attract immigrants who all have ebola. And that your longed-for barn conversion is actually a Russian missile silo capable of turning all of Chipping Sodbury into a nuclear desert for the next 10,000 years.

SUMMER

Let them eat soup

Ages ago someone told me that there is only one absolute rule in farming: the electric fence is always on. There is, however, another rule. Whatever you hope will happen, won't. And this year it's even worse than usual. Fertiliser prices are still sky high, wheat prices are plummeting, the weather is constantly bonkers and now we have Sir Starmer, who we hope won't be the next prime minister – meaning he will be – saying that he's going to start confiscating farmland so that it can be turned into 'sustainable, affordable, green housing for hard-working families in the low-income community'.

For two reasons, however, I'm completely unbothered. First, I've downloaded a new app. It's called Merlin and what it does is identify birdsong. This means I've been getting up at four every morning so that I can put names to all of those invisible participants in the dawn chorus.

It's incredible. There's a wall of squeaky noise and instantly it tells you you're listening to a Eurasian blackbird, a chaffinch, a goldfinch, a wren, a house sparrow and a dunnock. With a smattering of wood pigeon thrown in for good measure.

The newspapers arrive with more grim news for farmers, but by this stage I'm in the garden, going, 'It's picked out a golden oriole. What the hell's that doing in Oxfordshire?' Then Kaleb drops round to say he was hoping to do some spraying, which means it's too windy and he can't. But I'm not listening because my phone has detected a common raven. And a white wagtail. It's all I do these days and it's making me very happy.

The other reason why I'm not bothered about all the hiccups and Starmerisms is that I've had a brainwave, which, unlike the wheat, oats, oilseed rape, barley, echium, pigs, cows, sheep and goats, will secure the farm's financial future, no matter what the weather gods or the fools in Westminster do. I'm launching a new soup, made from nettles.

Unlike any other plant in the big wide world of farming, nettles do not need to be planted, or fertilised, or nurtured in any way. Even if you do absolutely nothing at all they will come and they will multiply and they will flourish. So they are completely free – and they are delicious.

The recipe for the new soup was developed by a young racing driver who works at the Diddly Squat farm shop, and it's a secret. But I've been feeding my new invention to guests these past few weeks and all of them say the same thing: 'That is gorgeous and it really does taste of nettles.' I find this baffling because how do they know what nettles taste like? The only person whom I've ever seen eat one is Chris Packham, so he is qualified to identify them using nothing but his mouth. But for everyone else it's the same as saying, 'Mmm, this tastes of battery acid.'

Buoyed by the positive reaction, I made a business plan and worked out that I could sell 500ml portions, chilled and packaged, for £5.80. Which is a bit weird because the main ingredient, apart from the cream, and the potatoes, and the chicken stock, and the butter, cost me nothing at all. How supermarkets can sell their soups for £2 less – when their ingredients have to be farmed – is beyond me.

Some people have said that no one will pay £5.80 to put a handful of nettles in their mouths but I think they will. So I decided to put my plan into action. This meant putting my Merlin bird app down for a few minutes and thinking of a way that nettles could be harvested.

Initially I deployed my Henry vacuum cleaner, which had worked so well at blackberry time, but it was a complete disaster. So I had to put the bird app down again and have a rethink. And soon I came up with the idea of using the same machines they use for harvesting tea.

The first was suspended from my shoulders in much the same way as Sigourney Weaver's big gun in *Aliens*. And it worked brilliantly. In just three minutes I had a cupful of only the tenderest top leaves but I'm afraid I couldn't continue because my back was in agony. So I switched to a much larger device that you push through the nettles like a hospital stretcher. That has a hedge cutter underneath it. And a fan for blowing the severed leaves into a big bag.

This didn't hurt my back at all, and it was hoovering up all the tender top leaves beautifully. And the not so tender bottom leaves. And the stalks and everything else that you find on a forest floor. Twigs. Dead mice. Deer faeces. There was a little bit of everything in there and Lisa thought this could be a marketing problem. Because while I might be able to get people to pay to put nettles in their mouths, they will almost certainly shy away if the label says the soup may contain trace elements of Bambi turd.

I was forced, eventually, to concur and came up with another harvesting idea. I'd use a combination of child labour and the minimum wage. It'll add slightly to the cost and eat into my margins, but as I write there's an army of kids in the woods playing what they call music and picking away merrily. Some of them even have gloves. There's also a skylark that my phone has just heard.

And me? Well, I'm sitting here hoping that my new

nettle soup becomes a firm farm shop favourite. Which means, of course, it'll be a disaster. But if it isn't, and I really have come up with a way of monetising the plague of nettles that choke the woods every summer, it could open the floodgates, because there's watercress in there as well, and wild garlic, and crayfish, and mushrooms. And all of these things can be turned into soup. I could become the next Baxter.

There are also a lot of deers. I'm not sure they can be turned into soup but they can definitely be turned into venison, which is free – and free-range – meat that's very healthy. I know it's popular at the dining tables in Scottish castles, but can ordinary people be persuaded to eat Bambi? And if they can, can I be persuaded to go out there with a gun and shoot one?

I'm not sure about that but I am sure of this. The 500 acres that we farm here using tractors and seed drills and so on are extremely unlikely to be profitable this year. But the 500 acres we don't farm at all could well be a gold mine.

I've even thought about using our most vigorous stream to produce electricity. Clean and free power to keep those permanently on electric fences permanently on. That didn't work, though, because the environment agency said no. I don't know why either. Probably because I was hoping they'd say yes.

In praise of goats

A friend called last week to say he'd watched my farming show and was thinking of maybe getting a few sheep for the paddocks at his weekend cottage in Wiltshire. To which I replied, 'Well, you obviously haven't watched my farming programme or you wouldn't be thinking of getting some sheeps for your weekend paddocks.'

I can see what he's thinking of course. He has the cottage and the cottage garden and every weekend he puts on his salmon pink trousers and wanders around with a pair of gifty-wifty secateurs. And he thinks this rural idyll would be enhanced if he had some sheeps to baa him to sleep at night.

Ha. First things first. We have to assume he won't eat his sheep or sell them to be converted into meat, which means he plans to keep them as pets. And that's idiotic because sheep are not like dogs. They won't lie by the fire, gazing at you adoringly. And they're not like cats either. They won't strut about on the dining room table with their tails raised like car aerials so that you can have a better view of their tea-towel holders. They aren't even like horses because you can't ride around on them.

As I've explained on these pages before, sheep want only one thing out of life and that is to end it. They will spend all day hunting around in their field for the most unusual and revolting way of achieving this and if nothing comes to hand they will simply stand there and rot. Or they will escape and get on the nearest main road so they can experience the sheer unbridled joy of being hit by an articulated lorry and bursting.

All of this is just about worthwhile if you have a sheep farming business, but if you just want some animals to make you feel rural and bucolic, almost everything is better than a sheep. I'd rather have a saltwater crocodile.

'Pigs, then?' he said. This is a much better idea. Pigs are fantastic and if you select a rare breed like my Oxford Sandy and Blacks, you are helping to maintain some of the nation's heritage. Plus, if you can get them to mate you will end up with piglets and I know of no creature that's more adorable.

But there are drawbacks. Pigs have a lot of nipples. Often as many as 14. Which means they can and do produce huge litters. So you start off with, say, five pigs and pretty soon you'll have more than a hundred. And then what are you going to do? Buy more land? Break out the apple sauce?

There's something else too. Pigs make the most godawful mess. In about a week your pretty paddocks will look like the Donbas. And the smell is so pungent that pretty soon you're going to get a letter from the parish council inviting you to bugger off back to London.

Cows are less prolific, which means you won't have a sudden population explosion to worry about. And they don't like to commit suicide or escape. And they keep the grass down. But I dunno. They are quite large, which means they are expensive. A sheep costs about £125; a pig is maybe £60. But a decent cow is more than £1,500. And also, because they are big, I find them quite frightening.

Most of the time they pay you no mind, but on some days it's like they've been on the vodka and Red Bull. They bounce up and down and lash out with their feets and it can be scary, sharing a small field with four tonnes of Tiggerish bouncy muscle. Best, really, to leave them to the professionals.

On the face of it, donkeys make a deal more sense. You can get them from a rescue centre and every day you can give them a carrot or an apple. You won't feel the need to ride around on them, your children will love them and they can become like outdoor tiny-maintenance pets. But. A word of warning. I used to keep donkeys and I learnt through bitter experience that when people talk about 'donkey's years', what they mean is 'eight years'. So, just after they become part of the family, they die.

Today I'm into goats. I just bought a fleet of 30 for a tenner each and so far things are going fairly well. They are not pretty little pygmies or some kind of rare breed that will produce jackets that Lisa can wear to Ascot. They're just run-of-the-mill man-goats that I've bought simply to keep the brambles at bay.

I was advised by Cheerful Charlie that they are psychopaths, and he's not alone in thinking this way. My dogs, for some reason, are terrified of them and I too have a healthy respect, as they will often head-butt me in the testes. There's more too. In the occult Satan is often seen to have a goat's head. According to the world's Bible enthusiasts this is because God created all the world's creatures, leaving Satan just one — the goat. And then God sent his wolves to tear the goats to pieces and that annoyed Satan, so, for some reason, he put out all of the goats' eyes and replaced them with his own. Which is why goats have rectangular pupils.

Actually they have rectangular pupils to broaden their field of vision and to see better at night, and they need to see well at night because that's when they do their planning. Unlike sheep, goats don't want to escape so they can be splattered all over the radiator grille of a

16-litre S-series Scania. They want their freedom so they can eat whatever's growing in the next field.

And we are talking here about animals that put Dickie Attenborough and Gordon Jackson to shame. Goats will move oil drums and feeding troughs nearer to fences so they can be used as launchpads. They will dig tunnels. They will build vaulting horses. And if you leave the right equipment lying around, they will make a glider and use that.

Not mine, though. To keep them in place I have two layers of electric fencing. The first delivers a mild shock, and if they persevere the second gives them a *Green Mile* kicking. I'd love to say it wasn't funny, watching them on day one as their noses got zapped. But it was. They actually say, 'Ow.' Just like I do. Only I then follow up by saying, 'Kaleb, you bastard! You said the fence was off.'

The fact is, though, that in two days they'd turned a rough bit of ground into what looked like a cross between a croquet lawn and a porn star's lady part. They hadn't escaped and they hadn't made a smell. And if you wear a cricket box when you're in there with them they are actually very playful.

So that, then, is my advice to anyone who wants a few animals for their weekend cottage paddocks. Goats. They're fun, cheap, easy to keep and, thanks to their rectangular eyes, they are excellent at keeping the vicar at bay.

OWL'S HOUSE

CARAVAN PARK

DRY STONE WALL

THE PIGS

ME

WASABI

MY DAM

DIDD

World According To
CLARKSON

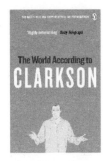
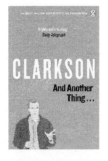
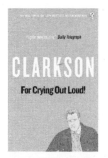
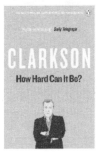
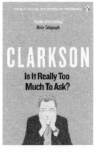
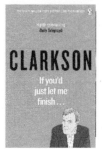

WHICH BOOK WILL YOU READ NEXT?

CLARKSON
ON CARS

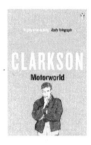
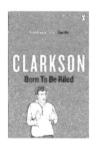
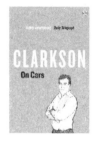
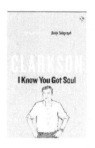

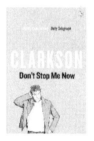
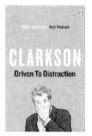
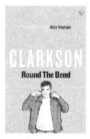
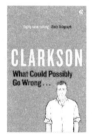

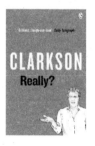

WHICH BOOK WILL YOU READ NEXT?

'Brilliant . . . laugh out loud'

Daily Telegraph

The No.1 *Sunday Times* bestseller

Jeremy Clarkson

Diddly Squat

A Year on the Farm

The No. 1 *Sunday Times*
Bestseller

NURTURING WRITERS SINCE 1935

'If you want a laugh, it's Jeremy Clarkson's
Diddly Squat: 'Til the Cows Come Home . . .'
Spectator

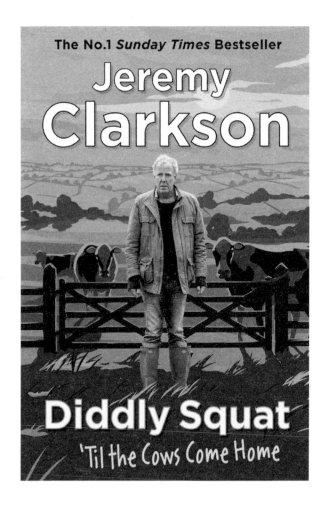

WHICH BOOK WILL YOU READ NEXT?